Principles and Standards for School Mathematics Navigations Series

NAVIGATING
through GEOMETRY
in
GRADES 9–12

Roger Day
Paul Kelley
Libby Krussel
Johnny W. Lott
James Hirstein

Johnny W. Lott
Grades 9–12 Editor

Peggy A. House
Navigations Series Editor

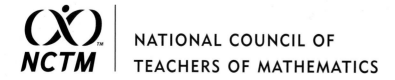

NATIONAL COUNCIL OF
TEACHERS OF MATHEMATICS

Library of Congress Cataloging-in-Publication Data:

Navigating through geometry in grades 9–12 / Roger Day … [et al.}; Johnny W. Lott, grades 9–12 editor.
  p. cm. — (Principles and standards for school mathematics navigations series)
Includes bibliographical references.
ISBN 0-87353-514-6
  1. Geometry—Study and teaching (Secondary)—United States. I. Day, Roger, 1958- II. Lott, Johnny W., 1944- III. Series.

QA461 .N3143 2001
516′.0071′273—dc21

                                                          2001055854

Dynamic Geometry is a registered trademark of Key Curriculum Press and is used with the permission of the trademark holder.

Printed in the United States of America

TABLE OF CONTENTS

Contents of CD-ROM

Applets

Tessellation Exploration (Chapter 1)
Viewing-Tube Geometry (Chapter 3)
Koch Snowflake Curve (Chapter 4)
Least Squares (Chapter 4)

Supplemental Materials

Time Line of Important Events in Map Making (Chapter 2)
Applications of the Global Positioning System (Chapter 2)
Trilateral Coordinates and Triangular Grids (Chapter 2)
Similar Triangles for "Scale Factors" (Chapter 3)
Directions for Sketchpad Users for "How Small Are the Squares?" (Chapter 4)

Blackline Masters

Readings from Publications of the National Council of Teachers of Mathematics

Godzilla: Fact or Fiction
Rick Billstein and James Trudnowski
NCTM Student Math Notes

Visual Thinking with Translations, Half-Turns, and Dilations
Tom Brieske
Mathematics Teacher

About This Book

Secondary school geometry has long served as a vehicle for teaching logical reasoning and the deductive method. In colleges and universities, geometry has commonly been approached as a branch of algebra or analysis. However, *Principles and Standards for School Mathematics* (National Council of Teachers of Mathematics [NCTM] 2000) sees geometry as affording slightly different and richer opportunities: "Geometry offers a means of describing, analyzing, and understanding the world and seeing beauty in its structures. Geometric ideas can be useful both in other areas of mathematics and in applied settings" (p. 309). Furthermore, "high school students should develop facility with a broad range of ways of representing geometric ideas—including coordinates, networks, transformations, vectors, and matrices—that allow multiple approaches to geometric problems that connect geometric interpretations to other contexts" (p. 309). Thus, *Principles and Standards* neither excludes from high school geometry the teaching of logical reasoning and the deductive method nor implies that geometry could not be taught as a branch of algebra or analysis, but it does expand the roles of geometric ideas beyond the traditional ones.

In a book as brief as this one, it is impossible to include all geometric topics and contexts that the authors or others deem important. We have had to choose carefully among topics in order to concentrate on overarching ideas from *Principles and Standards* that extend to other mathematical contexts. These include transformations, coordinates, and matrices, as well as the use of geometry to describe, analyze, and understand location in the world by means of the Global Positioning System (GPS). We also consider such traditional topics as congruence and similarity in the context of twenty-first-century developments, and we explore applications in which geometry is a tool for investigating the infinite.

We approach geometry through a transformational lens. Transformations provide the study of geometry with a functional basis that lends itself to extensions to algebra, statistics, and calculus. Transformations equip students for future explorations of computer science through the use of matrices and mappings. In applying transformations, we do not neglect the rich traditional geometry of past centuries but focus instead on an application from the centuries-old tradition of map making.

In this book, we assume that middle school students have "explored and discovered relationships among geometric shapes, often using dynamic geometry software" (NCTM 2000, p. 309). Moreover, we expect that middle school students will have begun to develop logical arguments that will underpin more formal reasoning and proof. We expect students to continue that development, calling on them to present justified arguments to support geometric claims.

Each chapter features a group of activities that take students through related geometry tasks. For each activity, we include goals, materials and equipment, and a "Discussion of the Activity" that briefly explains what students will be doing and learning. The activities themselves appear in reproducible blackline masters in the appendix, which also

A transformation is a one-to-one correspondence that maps the plane to itself.

Key to Icons

Principles and Standards

CD-ROM

Blackline Master

includes solutions. The activity pages move students step by step through a variety of concrete tasks, and "Discussion and Extension" questions call on them to reflect on what they have just done, often urging them to generalize from their results. The blackline masters of the activities also appear on the CD-ROM that accompanies this book.

Three different icons appear in the book, as shown here in the key. One alerts readers to relevant material in *Principles and Standards for School Mathematics*, another points them to supplementary materials on the CD-ROM, and a third signals the blackline masters and indicates their locations in the appendix.

As you and your students work through the activities, you will sometimes need access to Dynamic Geometry® software or a calculator or computer with software that handles data and allows you to produce geometric images and graphs. The authors believe that all teachers need to adopt the Technology Principle from *Principles and Standards*: "Technology is essential in teaching and learning mathematics; it influences the mathematics that is taught and enhances students' learning" (NCTM 2000, p. 24). The activities that use technology make note of this requirement in their lists of materials. Some of them could be approached without the technology, but we believe that other approaches would not be likely to be as effective.

As you read, think about what a geometry course for students in today's world could and should be if it were built around the Principles and Standards. We have tried to provide you with food for thought as well as to give you activities to challenge your students and enhance your own professional development.

We would like to give particular thanks for the special contributions of the following people whose ideas and input have added immeasurably to the variety and scope of the book:

Collin Joyce

Tami Martin

Corey Andreasen

Paul Anderla

Sara Lenertz

Charles Vonder Embse

NAVIGATIONS SERIES

GRADES 9–12

NAVIGATING *through* GEOMETRY

Introduction

Both in the development of mathematics by ancient civilizations and in the intellectual development of individual human beings, the spatial and geometric properties of the physical environment are among the first mathematical ideas to emerge. Geometry enables us to describe, analyze, and understand our physical world, so there is little wonder that it holds a central place in mathematics or that it should be a focus throughout the school mathematics curriculum.

When very young children begin school, they already possess many rudimentary concepts of shape and space that form the foundation for the geometric knowledge and spatial reasoning that should develop throughout the years. *Principles and Standards for School Mathematics* (National Council of Teachers of Mathematics [NCTM] 2000) recognizes the importance of a strong focus on geometry throughout the entire prekindergarten–grade 12 curriculum, a focus that emphasizes learning to—

- analyze characteristics and properties of two- and three-dimensional geometric shapes and develop mathematical arguments about geometric relationships;
- specify locations and describe spatial relationships using coordinate geometry and other representational systems;
- apply transformations and use symmetry to analyze mathematical situations;
- use visualization, spatial reasoning, and geometric modeling to solve problems. (P. 41)

Geometry not only provides a means for describing, analyzing, and understanding structures in the world around us but also introduces an experience of mathematics that complements and supports the study of other aspects of mathematics such as number and measurement. Geometry offers powerful tools for representing and solving problems in all areas of mathematics, in other school subjects, and in everyday applications. *Principles and Standards* presents a vision of how geometric concepts and reasoning should develop and deepen over the course of the school mathematics curriculum. The *Navigating through Geometry* books elaborate that vision by showing how important geometric concepts can be introduced, how they grow, what to expect of students during and at the end of each grade band, how to assess what students know, and how representative instructional activities can help translate the vision of *Principles and Standards* into classroom practice and student learning.

Foundational Components of Geometric Thinking

The Geometry Standard emphasizes as major unifying ideas *shape* and the ability to analyze characteristics and properties of two- and three-dimensional objects and develop mathematical arguments about geometric relationships; *location* and the ability to specify positions and describe spatial relationships using various representational systems; *transformations* and the ability to apply motions, symmetry, and scaling to analyze mathematical situations; and *visualization* and the ability to create and manipulate mental images and apply spatial reasoning and geometric modeling to solve problems. Each of these components of geometric thinking requires nurturing and developing throughout the school curriculum.

Analyzing characteristics and properties of shapes

By the time the youngest children begin formal schooling, they have already formed many concepts of shape, although their understanding is largely at the level of recognizing shapes by their general appearance and they frequently describe shapes in terms of familiar objects such as a box or a ball. In the primary grades, children should have ample opportunities to refine and focus their understanding and to gradually develop a mathematical vocabulary. They also should learn to recognize and name the parts of two- and three-dimensional shapes, such as the sides and the "corners," or vertices. Teachers should provide frequent hands-on experiences with materials, including technology, that help the students focus on attributes of various shapes, such as that a square is a special rectangle with all four sides the same length or that pyramids always have triangular faces that meet at a common point. Experiences that promote such outcomes include building and drawing shapes; comparing shapes and describing how they are alike and how they are different; sorting shapes according to one or more attributes; cutting or separating shapes into component parts and reassembling the parts to form the original or different shapes; and identifying shapes found in everyday objects or in the classroom, home, or neighborhood. Throughout such activities, teachers must take care to ensure that the children encounter both examples and nonexamples of common shapes and that they see those examples in many different contexts and orientations so that they learn to identify a triangle or a rectangle, for example, no matter what material it is made of or how it is positioned in space.

As children progress to the higher elementary grades, they should continue to identify, compare, classify, and analyze increasingly more complex two- and three-dimensional shapes, and they should expand their mathematical vocabulary and refine their ability to describe shapes and their attributes. As they do so, they begin to develop generalizations about classes of shapes, such as prisms or parallelograms, and to formulate definitions for those classes. They also include in their study not only two- and three-dimensional shapes but points, lines, angles, and more-precise relationships such as parallelism and perpendicularity. They begin to explore properties of area and perimeter and to pose questions related to those measurement concepts; they might, for example, use tangram pieces to investigate whether shapes that are different, such as a rectangle, a trapezoid, and a nonrectangular parallelogram, can have the same area. They also develop and explore concepts of congruence and similarity, which they express in terms of shapes that "match exactly" (congruence) or shapes that "look alike" except for "magnifying" or "shrinking" (similarity). In grades 3–5, there should be a growing emphasis on making conjectures about geometric properties and relationships and formulating mathematical arguments to substantiate or refute those conjectures; for example, students might use tiles or grid paper to show that whenever the sides of one square are twice as long as the sides of another square, then four of the smaller squares will "fit inside" or "cover" the larger square, or they might measure to demonstrate that a rectangle, trapezoid, and nonrectangular parallelogram that have equal area do not necessarily have the same perimeter.

The informal knowledge and intuitive notions developed in the elementary grades receive more-careful examination and more-precise description in the middle grades. Descriptions, definitions, and classification schemes take account of multiple properties, such as lengths, perimeters, areas, volumes, and angle measures, and students should use those characteristics to analyze more-sophisticated relationships by, for instance, developing a classification scheme for quadrilaterals that accurately represents some classes of quadrilaterals (e.g., squares) as special cases or subsets of other classes (e.g., rectangles or rhombuses). At the same time, they should develop the more precise language needed to communicate ideas such as that all squares are rectangles but not all rectangles are squares. Students in the middle grades should also investigate what properties of certain shapes are necessary and adequate to define the class; they might explore, for example, the following question: Among the many characteristics of rhombuses, including congruent sides, opposite sides parallel, opposite angles congruent, diagonals that bisect each other, and perpendicular diagonals, which characteristics can be used to define rhombuses and to differentiate them from all other quadrilaterals? In a similar manner, other concepts introduced informally in the lower grades, including *congruence* and *similarity*, should be established more precisely and quantitatively during the middle grades, and special geometric relationships, including the Pythagorean relationship and formulas for determining the perimeter, area, and volume of various shapes, should be developed and applied. All these explorations should be carried out with the aid of hands-on materials and dynamic geometry software, and all should be conducted in an environment in which students are expected and encouraged to make and test conjectures and develop convincing arguments, based on both inductive and deductive reasoning, to justify their conclusions.

By the time students reach high school, they should be able to extend and apply the geometric knowledge developed earlier to establish or refute conjectures, deduce new knowledge from previously established facts, and solve geometric problems. They should be helped to extend the knowledge gained from specific problems or cases to more-general classes of objects and thus to establish the validity of geometric conjectures, prove theorems, and critique arguments proposed by others. As they do so, students should organize their knowledge systematically in order to understand the role of definitions and axioms and to appreciate the connectedness of logical chains, recognizing, for example, that if a result is proved true for an arbitrary parallelogram, then it automatically applies to all rectangles and rhombuses.

Specifying locations and describing spatial relationships

The importance of location and spatial relationships becomes apparent when we try to answer questions such as Where is it? (location), How far is it? (distance), Which way is it? (direction), and How is it oriented? (position). Typically, the first answers that children give to questions such as these are in relation to other objects: on the chair, next to the book, under the bed. In the primary grades, teachers help students develop a sense of location and spatial relationship by developing those early ideas. Using physical objects, often to illustrate stories, or physically acting out a relationship, children learn the meaning of such concepts as above, below, in front, behind, between, to the left, to the right, next to, and other relative positions. In time they add concepts of distance and direction, such as three steps forward, and they learn to combine such descriptions to lay out routes (e.g., walk to the door, turn left, go to the end of the corridor). Students begin to represent such physical notions of location, distance, and space both as verbal instructions and as diagrams or maps, and they learn to follow verbal directives and to read maps as means to locating a hidden object or reaching a desired destination. As their skills in representing locations increase, students should add more quantitative details by, for instance, pacing off or measuring distances to better communicate "how far" or adding a simple coordinate system to define a location more precisely.

In grades 3–5, students' understanding of location, direction, and distance are applied to increasingly more complex situations. They become more precise in their measurements and begin to examine situations to determine whether there is more than one route between two points or if there is a shortest distance between them. During these years, students should come to recognize that some positional representations are relative (e.g., *left* or *right*) or subjective (e.g., *near* or *far*), whereas others are fixed (e.g., *north* or *west*) and unambiguous (e.g., *between*) and that directions are not always interchangeable (e.g., two blocks north, then three blocks west does not take you to the same destination as two blocks west, then three blocks north, but three blocks west, then two blocks north does have the same end point as two blocks north, then three blocks west). They also should become more attentive to orientation; they might determine the direction that an object faces, whether it has been reflected or rotated from its initial position, or the distance and direction that it has been moved, all of which are closely related to ideas of transformations discussed later. It is particularly convenient and appropriate for students to explore the concepts of location and position by using grids together with graphical representations, physical models, and computer programs; as they continue their explorations on a grid, students also should learn to specify ordered pairs of numbers to represent coordinates and to use coordinates in locating points, describing paths, and determining distances along grid lines.

In the middle grades, the ideas established in elementary school should continue to be developed, and in addition, geometric ideas of location and distance can be linked to developing algebraic concepts as students apply coordinate geometry to the study of shapes and relationships. For example, the study of linear functions in algebra is related to the determination of the slopes of the line segments that form the sides of polygons, and these values in turn are used to determine relationships such as parallelism or perpendicularity of sides, which are used to analyze and classify the polygons; the Pythagorean relationship is applied to the coordinate plane to establish a method of determining the distance between points or the lengths of segments; and coordinates can be used to locate the midpoints of segments.

High school students should extend the geometric concepts of Cartesian coordinates used in lower grades to other coordinate systems, including polar, spherical, or navigational systems, and use them to analyze geometric situations. They also should develop facility in translating between different coordinate representations and should understand that each representation offers certain advantages in specific situations. During the secondary school years, students also learn to apply trigonometric relationships to solve problems involving location, distance, direction, and position, and analytical methods continue to be used, further strengthening the connection between algebra and geometry.

Applying transformations and symmetry

Among the early geometric discoveries that children make is that shapes can be moved without being changed: a triangle is still the same triangle even if it is flipped over or slid across the table, and a puzzle piece may need to be turned in order for it to fit into the desired space. Such intuitions are the starting point for studying transformations when children enter school. This important aspect of spatial learning in the primary school years engages students in exploring the motions of slides, flips, and turns, which leads to the discovery that such motions alter an object's location or orientation but not its size or shape. Primary-grades teachers should guide children to look for, describe, and explore symmetric shapes, which they can do informally by folding paper, tracing, creating designs with tiles, and investigating reflections in mirrors. Explorations that children enjoy in the primary grades, such as folding the net of a rectangular prism to make a "jacket" for a block, also involve relationships between two- and three-dimensional shapes.

As children move into the upper elementary grades, the more informal notions of slides, flips, and turns are treated with greater precision as translations, reflections, and rotations, and attention is directed to what parameters must be specified in order to describe those transformations (e.g., slide [translate] the square ten centimeters to the right; flip [reflect] the triangle over its hypotenuse; turn [rotate] the drawing a quarter-turn clockwise). Students also learn that transformations can be used to demonstrate that two shapes are congruent if one can be moved so that it exactly coincides with the other. They should then be helped to extend that notion by being challenged to visualize and mentally manipulate shapes, describing mathematically a series of motions that can be used to demonstrate congruence or predicting the result of certain transformations before actually performing them with physical objects or symbolic representations. Such growing precision extends as well to symmetry as students learn to specify all the reflection lines or the center and the degrees of rotation in a symmetric figure or design.

In grades 6–8, transformation geometry can be a powerful tool for exploring spatial and geometric ideas. Not only are rigid transformations used to deepen students' understanding of such concepts as congruence, symmetry, and the properties of polygons, but dilations and the notions of scaling and similarity, which are closely linked to proportional reasoning, are introduced in the middle grades. Additionally, in their study of transformations, middle-grades students should, for instance, generalize the result of two successive reflections over parallel lines and compare that outcome to the result of two successive reflections over intersecting lines. They should also begin to quantify and formalize aspects of transformations, establishing, for example, that in a reflection each point on the original object is the same distance from the mirror line as the corresponding point on the image. Physical manipulatives, such as mirrors or other reflective devices, and dynamic geometry software are especially useful in conducting such investigations.

In high school, the study of transformations can be further enriched by the use of function notation, coordinates, vectors, and matrices to describe and investigate transformations, including both isometries and dilations. Students should develop certain basic "tools" such as determining a matrix representation for accomplishing a reflection over the line $y = x$ or other common transformations, and they should relate the composition of transformations to matrix multiplication and apply those concepts to the solution of problems.

Using visualization, spatial reasoning, and geometric modeling

The ability to create mental images of two- and three-dimensional objects, to visualize how objects appear from different perspectives, to formulate representations of how objects are positioned relative to other objects, to relate two-dimensional renderings to the three-dimensional objects that they represent, to predict how appearances will vary as the result of one or more transformations, and to create spatial representations to model various mathematical situations are among the most important outcomes of the study of geometry.

Young children begin to develop their spatial visualization by initially manipulating physical objects and later extending their manipulations to mental images. Teachers in the primary grades may help children develop spatial memory and spatial visualization by asking them to recall and describe hidden objects or by having them describe how an object would look if viewed from a different side. They may ask children to imagine, and later explore and verify, what will happen when a given shape is cut in two in a certain way or to predict and demonstrate what other shapes could result if that same object were cut in a different manner. Children should also experiment with different shapes and formulate descriptions of them, perhaps by creating a shape from tangrams and taking turns to describe what each one sees in the figure. Students should also learn to read and draw simple maps and to give and follow directions—for example, giving a classmate verbal instructions for going from the classroom to the cafeteria. Opportunities abound to develop spatial visualization in connection with other topics and subjects by, for instance, demonstrating that even numbers can always—whereas odd numbers can never—be arranged in two equal rows or highlighting spatial concepts during art or physical education lessons. Children should have ample opportunity to discover that spatial reasoning and geometric modeling can contribute to understanding and solving a wide variety of problems that involve number, data, and measurement and that have numerous applications.

As students move through grades 3–5, they become more adept at reasoning about spatial properties and relationships among shapes; they might develop strategies to calculate the area of a garden plot by subdividing it into component rectangles or relate the area of a trapezoid or a parallelogram to the area of the rectangle that is formed by cutting and reassembling the original quadrilateral. Relating three-dimensional shapes to their two-dimensional representations becomes an important topic in these grades as students discover how to build three-dimensional objects from two-dimensional representations, and vice versa; construct and fold nets of solids; examine diagrams of nets to predict which ones can or cannot be folded to form a certain prism; or mentally manipulate a shape to produce an accurate picture of hidden parts. Applying geometric reasoning and modeling to solve problems in all areas of mathematics, as well as in other contexts, should continue to be a principal focus of the curriculum.

The skills of spatial visualization and geometric reasoning that emerge in the lower grades should become more refined and sophisticated in grades 6–8 as students solve problems involving distance, area, volume, surface area, angle measure, and other quantifiable properties. Students should be guided to develop, understand, and apply important formulas for calculating the length, area, or volume of selected shapes. As they explore relationships using physical models and appropriate technology, students should begin to establish and give arguments to support important generalizations; they might, for example, demonstrate why, when the side of a cube is tripled, the surface area of the enlarged cube is nine times the surface area of the original whereas the volume of the enlargement is twenty-seven times that of the original. Geometric models for algebraic and numerical relationships help students integrate important concepts from all strands of the mathematics curriculum; manipulatives and computer programs that connect geometric, algebraic, and numerical concepts contribute to students' developing mathematical maturity and enable them to solve more-complex problems both within mathematics and in other subjects.

As students progress through high school, their visualization skills should extend from representations on the familiar two- and three-dimensional rectangular coordinate systems to analogous representations on a spherical surface or in a spherical space; investigations that connected two-dimensional representations of polyhedra with three-dimensional representations of them later evolve into challenges of projecting a spherical surface onto a plane and producing a two-dimensional map of a three-dimensional surface. Producing perspective drawings, visualizing the resultant cross section when a plane slices a solid object, predicting the three-dimensional shape that results when a plane figure is swept 360 degrees about an axis, and navigating in a spherical frame of reference are examples of spatial ideas that should evolve as students progress through school. In high school, too, geometric representations can be of great benefit when studying topics involving algebra, measurement, number, and data, and the application of geometric ideas to the solution of problems across mathematics and in other disciplines is one of the major goals of the curriculum.

Developing a Geometry Curriculum

A curriculum that fosters the development of geometric thinking envisioned in *Principles and Standards* must be coherent, developmental, focused, and well articulated, not simply a collection of lessons or activities. Geometric ideas should be introduced in the earliest years of schooling and then must deepen and expand as students progress through the grades. As they move through school, children should receive instruction that links to, and builds on, the foundation of earlier years; they must continually be challenged to apply increasingly more sophisticated spatial thinking to solve problems in all areas of mathematics as well as in other school, home, and life situations.

These Navigations books do not attempt to describe a complete geometry curriculum. Rather, the four *Navigating through Geometry* books illustrate how selected "big ideas" of geometry develop across the prekindergarten–grade 12 curriculum. Many of the concepts presented in these geometry books will be encountered again in other contexts related to the Algebra, Number, Measurement, and Data Standards; in the Navigations books, as in the classroom, the concepts described under the Geometry Standard reinforce and enhance students' understanding of the other strands.

Geometry is essential to the vision of mathematics education set forth in *Principles and Standards for School Mathematics* because the methods and ideas of geometry are indispensable components of mathematical literacy. The *Navigating through Geometry* books are offered as guides to help educators set a course for successful implementation of the very important Geometry Standard.

NAVIGATIONS
SERIES

GRADES 9–12

NAVIGATING *through* GEOMETRY

Chapter 1
Transforming Our World

Geometry has played a vital role in mathematics for centuries. It formed the basis for much of the study of mathematics in the past and continues to allow students to explore topics that extend to other branches of mathematics and other disciplines. To explore the visual appeal of geometry and to think about the usefulness of the subject to engineers, scientists, artists, mathematicians, and others, we set the stage in this chapter for a systematic approach to geometry using transformations. *Principles and Standards for School Mathematics* (NCTM 2000) states that students in grades 9–12 should be able to "apply transformations and use symmetry to analyze mathematical situations" (p. 308). The initial, distance-preserving transformations (isometries) used by high school students to examine geometric ideas are translations, reflections, rotations, and glide reflections, as shown in table 1.1.

As teachers acquaint their students with the information in the table, they may wish to ask students why it is customary for studies of the properties of transformations to consider what happens to just three non-collinear points of the plane (or the triangle determined by those three points). To help students respond, teachers might ask them how a plane is determined. If they know what will happen to three noncollinear points, can they determine what will happen to other points of the plane?

Two of the expectations for students in grades 9–12 expressed in *Principles and Standards* are that they be able to—

• understand and represent translations, reflections, rotations, and dilations [discussed in chapter 3 of the current book] of objects in the plane by using sketches, coordinates, vectors, function notation, and matrices;

Table 1.1

Representations of reflection, translation, rotation, and glide reflection

Type of Transformation	Representation	Properties
Reflection in line *m*	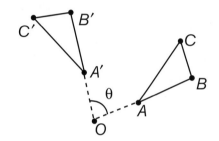	Any point *A* on *m* is its own image. For any point *C* not on *m*, *m* is the perpendicular bisector of $\overline{CC'}$ where *C'* is the image of *C*. We write $R_m(\triangle ABC) = \triangle A'B'C'$.
Rotation with center *O* and with measure θ	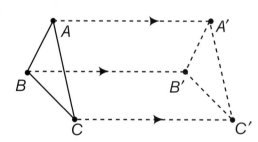	Each point *A* is rotated θ° on a circle whose center is *O* and whose radius is *OA*. (If θ is positive, the rotation is counter-clockwise; if θ is negative, then the rotation is clockwise.) We write $R_{O,\theta}(\triangle ABC) = \triangle A'B'C'$.
Translation by vector $\overrightarrow{AA'}$	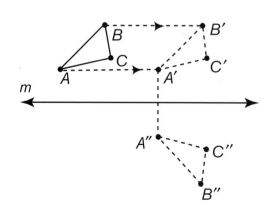	Each point *A* is moved along a vector $\overrightarrow{AA'}$ so that $AA' = BB' = CC'$ and $\overleftrightarrow{AA'} \parallel \overleftrightarrow{BB'} \parallel \overleftrightarrow{CC'}$. We write $T_{A \to A'}(\triangle ABC) = \triangle A'B'C'$.
Glide reflection with translation by vector $\overrightarrow{AA'}$ and reflection in line *m*		A point *A* is translated along a vector $\overrightarrow{AA'}$ and then reflected in line *m*, where $\overleftrightarrow{AA'} \parallel m$. We write the glide reflection as a composition of functions as $$R_m(T_{A \to A'}\triangle ABC) = R_m(\triangle A'B'C')$$ $$= \triangle A''B''C''.$$

- use various representations to help understand the effects of simple transformations and their compositions. (NCTM 2000, p. 308)

To understand and represent transformations mathematically, students need to see and use the necessary mathematical objects. This requirement does not preclude students from seeing the geometry in a real-world setting but demands that they work with different representations of the objects in order to understand what they are seeing from a mathematical point of view. To use transformations to analyze a mathematical situation, consider the wallpaper sample in figure 1.1.

If the shading in the figure is ignored, transformations can be used to create any bird from any other bird. For example, the bird marked *A* can be translated to obtain the bird marked *A'*, and the bird marked *B* can be rotated to obtain the one marked *B'*.

A transformation approach to geometry provides a formal approach to Euclid's technique of moving one figure on top of another to determine congruence. The mathematical study of transformations was organized by Felix Klein (1849–1925) in the *Erlanger Programm*, in which he described geometry as the study of properties that remain unchanged (invariant) under the group of transformations of a plane (Eves 1969, pp. 187–88). Though transformations were systematized around the beginning of the twentieth century, they were not widely included in secondary school curricula for many years, until their appearance in *Modern Mathematics*, Volumes 1–2 (Papy 1968); *Motion Geometry*, Books 1–4, from the University of Illinois Committee on School Mathematics (Phillips and Zwoyer 1969); *Geometry: A Transformational Approach* (Coxford and Usiskin 1971); and *Geometry: Constructions and Transformations* (Dayoub and Lott 1977). Since the publication of these books, a transformation approach has appeared in many texts, such as those by Hoffer (1979) and Serra (1989).

Although reflections, translations, and rotations are the primary transformations considered in these texts and in *Principles and Standards*, examination of an additional function, the glide reflection, helps to complete the picture. To consider various representations of transformations in this chapter, we present activities that range from paper folding to matrix operations to geometric constructions using technology.

What Are the Transformations?

Reflections

A reflection is a primary building block for all isometries, and a variety of concrete methods can be used to study it. In middle school, a student may have constructed a reflection by folding paper or using a reflective device such as a Mira (a small plastic instrument that students can use to create a reflection of a figure and draw its image by looking through the plastic). This concrete approach is reviewed in the following activity, Fold Me! Flip Me!

Fig. **1.1.**

Wallpaper created by Tessellation Exploration.

Tessellation Exploration is available in a demonstration version on the CD-ROM, courtesy of Tom Snyder Productions.

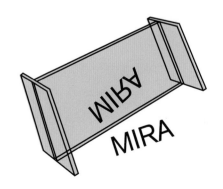

Fold Me! Flip Me!

Goals

- Draw and construct two-dimensional geometric objects using a variety of tools
- Understand and represent reflections
- Establish conjectures and proof by means of constructions

Materials and Equipment

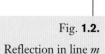

p. 78–79

- A copy of the activity pages for each student
- Paper for paper folding
- A Mira or an equivalent reflective device for each student

Discussion of the Activity

The activity shows that properties of a reflection, including the fact that the reflecting line is the perpendicular bisector of the segment connecting a point and its image, aid in the study of diverse topics in mathematics—such as line symmetry in a figure and the notion of an even function in algebra—as well as in archaeology. The activity directs students to use both paper folding and a reflective device to construct reflection images. Students should consider that, as shown in figure 1.2a, the reflecting line *m* is the perpendicular bisector of every segment connecting a point and its image under the reflection. Either folding paper or using a Mira (or similar device) will convince a student intuitively that this is true.

Other observations in the early stages of this activity are that (1) if any part of a figure is drawn on the same side of line *m* as *D*, then the image of that figure reflected in line *m* must be on the same side of *m* as *D'*; and (2) the *orientation* of a figure changes when it is reflected. That is, if the labels on a figure appear in the order *A*, *B*, *C* when they are read clockwise, then the labels on the image of that figure under a reflection will appear in the order *A'*, *C'*, *B'* when they are read clockwise as well.

An application of the fact that the reflecting line is the perpendicular bisector of a segment connecting a point and its image can be used to prove the following result: *Every point on an angle bisector is equidistant*

Fig. **1.2.**

Reflection in line *m*

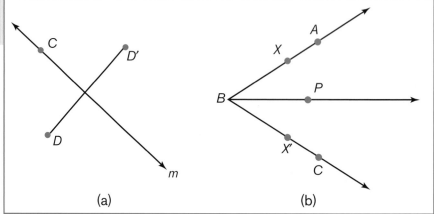

(a)   (b)

from the sides of the angle that it bisects. To see that this is true, consider figure 1.2b. Let $\overrightarrow{PX}$ be perpendicular to $\overrightarrow{BA}$, and let $\overrightarrow{PX'}$ be perpendicular to $\overrightarrow{BC}$. As a result, $\triangle X'BP$, the image of $\triangle XBP$ in the line containing $\overrightarrow{BP}$, is congruent to $\triangle XBP$. We know that this is so because the reflection preserves both angle size and distance. Therefore, the image of $\overrightarrow{BA}$ must be $\overrightarrow{BC}$; the image of $\angle ABP$ must be $\angle CBP$, and $\overrightarrow{BP}$ is its own image. Thus, the triangles are congruent by Side-Angle-Side (SAS), and $\overline{PX}$ and $\overline{PX'}$ must be the same length. Hence, any point P on the angle bisector must be equidistant from the sides of the bisected angle. Students should be able to argue in a similar manner that *every point on the perpendicular bisector of a segment is equidistant from the endpoints of the segment.*

Students use the last result when they try to locate the center of an ancient plate from the remaining pottery shard pictured in figure 1.3. This problem presents them with an application of reflection (or line) symmetry. Because a circle has infinitely many lines of symmetry and each line of symmetry can be considered a reflecting line that maps the circle onto itself, the lines of symmetry are the perpendicular bisectors of the chords of the circle, and all the lines contain the center of the circle. Thus, by finding the intersection of any two lines of symmetry of the circle that contains the shard, students can find the center of the circle and determine its radius.

Exploring Mirror Images

Determining the size of an image in a mirror is a different application of a reflection, this time from the realm of science. Figure 1.4 shows the character named Polygon from the Figure This! Web site (www.figurethis.org). Polygon is looking at her reflection in a mirror hung flat on a wall. If the mirror image depicts Polygon from head to foot, how much of herself can she see as she walks toward the mirror or away from it? To answer this question, we use a property of a reflection in a mirror: *The angle of incidence is equal to the angle of reflection.* This property, depicted in figure 1.5, is explored in the activity Mirror, Mirror, on the Wall.

Fig. **1.3.**

Pottery shard from a plate

Fig. **1.4.**

Figure This! character Polygon reflected in a mirror

Copyright © 2001 by Widmeyer Communications. Reprinted by permission.

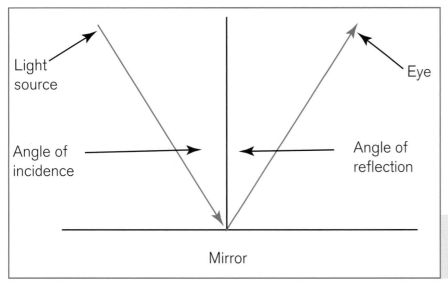

Fig. **1.5.**

Angle of incidence is congruent to the angle of reflection

Mirror, Mirror, on the Wall

Goals

- Understand properties of a reflection
- Establish conjectures and proof by means of constructions

Materials and Equipment

- A copy of the activity pages for each student
- A mirror for each student
- A straightedge (or Mira edge) for each student
- Geometry utility software

Discussion of the Activity

Using the measures of the angles of incidence and reflection helps students to answer the question about the size of the image that Polygon sees of herself when she walks away from or toward the mirror. Discussion and Extension question 3, about the size of a so-called full-length mirror that a store might stock, can be used to assess a student's ability to apply the knowledge of the properties of a reflection. Before teachers do this activity with their students, they may want to examine available "full-length" mirrors in local stores. They may also want to bring a sample mirror to the classroom to have students act out the activity.

p. 80

Two activities in chapter 4 extend the ideas developed in Mirror, Mirror, on the Wall. Smoke and Mirrors (see pp. 69 and 125) and Discontinuous, That's What You Are! (see pp. 70 and 126) engage students in investigations of mirror images formed by kaleidoscopes.

Geometric notions are clearly connected to the real world, as the archaeological and mirror examples show. They are also embedded in the mathematical world, as we have previously seen in the geometric proof of the fact that points on an angle bisector are equidistant from the sides of the angle. We can see this clearly in the algebraic graph in figure 1.6, as well.

Consider the function that the graph shows: $f(x) = \cos(x)$. This function maps the set of real numbers to the set $[-1, 1]$. The graph of that function is seen in figure 1.6. As shown, the part of the function's graph that lies to the right of the $f(x)$ axis is reflected in that axis to obtain the portion to the left of the axis. This function is an example of an *even function*—that is, a function such that $f(x) = f(-x)$. Students can be asked to name some other common algebraic graphs that have this geometric property.

Translations

An example of a translation appears in the wallpaper shown in figure 1.7. A single image of a frog can be slid, or *translated*, to obtain the other frogs in the pattern. The pictured translation, which takes point A to point A', could not be used to create the entire figure but only a series of frogs in a strip between two parallel lines. A simpler example is seen in figure 1.8. Once one strip of characters is created using the vector $\overrightarrow{AA'}$, the whole strip can be translated (with different translations) to create the entire wallpaper pattern in the plane.

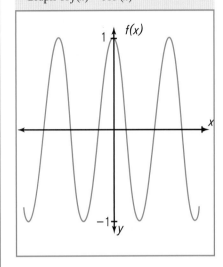

Fig. **1.6**

Graph of $f(x) = \cos(x)$

Fig. **1.7.**

Wallpaper translation design

To find students' wallpaper constructions on the Web, search on "tessellations" or "Tessellation Exploration."

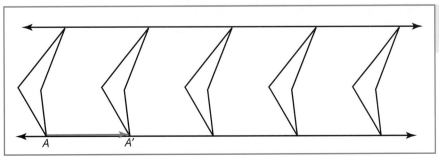

Fig. **1.8.**

Translation strip

Translations such as the one in figure 1.7 can be constructed with geometry utilities such as Geometer's Sketchpad, Cabri Geometry II, and Tessellation Exploration. In fact, Tessellation Exploration can be used to create the entire piece of wallpaper. Students can construct simple translations using tracing paper, as seen in figure 1.9. Such translations are typically undertaken before the high school level, and many Web sites show students' constructions.

Fig. **1.9.**

Translation constructed using tracing paper

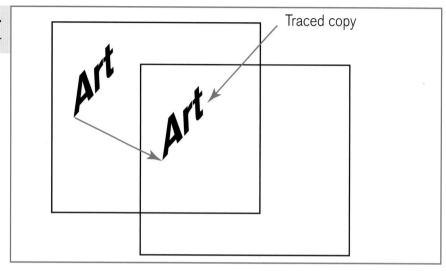

Several approaches can be used to study geometric translations formally. These include using vectors, considering a translation as a composition of two reflections, and extending both methods through the use of matrices. We have chosen the last two approaches in order to address the expectation of *Principles and Standards* that students in grades 9–12 will understand simple transformations and their compositions.

The activity Slide Me Now shows how a translation can be constructed by composing two reflections in parallel lines. In this activity, $\triangle ABC$ is reflected in line m to obtain the image $\triangle A'B'C'$. Then $\triangle A'B'C'$ is reflected in line n to obtain $\triangle A''B''C''$. Thus, the composition of reflections in lines m and n has the same result as the translation that takes A to A'', or

$$r_n \circ r_m (\triangle ABC) = \triangle A''B''C'' = T_{A \to A''}(\triangle ABC).$$

Among things that students should notice when studying a translation is that the orientation of the final image and that of the original are the same. Further, they should learn and be able to argue that the distance between the parallel reflecting lines is exactly half the length of the translation vector.

Several notations are used for translations, including the one shown here, $T_{A \to A''}$. Another description of a translation uses coordinates in expressions such as "translate by (5, 0)." This expression is used to signify that every point of a plane has 5 added to its x-coordinate and 0 to its y-coordinate.

Among other things that students should notice when studying a translation is that the orientation of the final image and that of the original are the same.

Slide Me Now

Goals

- Understand properties of a translation
- Construct a translation as the composition of reflections in parallel lines
- Establish conjectures and proof by means of constructions

Materials and Equipment

- A copy of the activity pages for each student
- A Mira or an equivalent reflective device for each student
- A protractor or ruler for each student

p. 82

Discussion of the Activity

Slide Me Now challenges students to construct a translation using two reflections in parallel lines as well as to determine two parallel lines that could be used to accomplish the translation by two reflections if the translation is given as a vector. The interplay among the different representations of a translation begins to demonstrate the power of integrated mathematics in considering a concept.

Teachers may want to have students argue that the image obtained when two reflections in parallel lines are composed is the same as a translation by a vector whose length is twice the distance between the two parallel lines and whose direction is from the first line to the second line.

Rotations

Recall from table 1.1 that a rotation is a transformation that fixes some point as the center and rotates every other point in the plane a specified number of degrees in a given direction (clockwise or counterclockwise). An example of a rotation appears in figure 1.10, which shows work by master designer and artist Scott Kim. The activity Design This investigates this art.

Fig. **1.10.**

Scott Kim's signature work (Copyright © 2000 by Scott Kim, www.scottkim.com). Reprinted by permission of the designer.

Design This

Goals

- Understand properties of a rotation
- Construct a rotation as the composition of reflections in intersecting lines
- Establish conjectures and proof by means of constructions

Materials and Equipment

p. 84

- A copy of the activity page for each student
- A Mira or an equivalent reflective device for each student
- Tracing paper for each student
- Pushpin for each student

Discussion of the Activity

Design This encourages students to trace Scott Kim's design (or a portion of it) to see if they can find a *turn center*—that is, a point that will allow it to be rotated onto itself in such a way that the tracing matches the original. After that, teachers may want to ask the students about the relation of the original design to the final image when a design is reflected first in the *x*-axis of a graph and then the resulting image is reflected in the *y*-axis. Thinking about such questions can serve as a prelude to reflecting a design in one line to find its image and then reflecting that image in a line that intersects the first, to find a final image. The resulting image will be a rotation image of the original and will have the same orientation. The center of the rotation will be the point of intersection of the two lines, and the measure of the turn angle will be twice the measure of one of the angles formed by the intersecting lines. Asking students to think about the reflection lines and to consider the order in which they performed the reflections should provide a clue about whether the turn is clockwise or counterclockwise.

Scott Kim's signature artwork has *turn symmetry* of 180° about its center. Teachers can ask students questions such as the following in order to build on and extend their appreciation of this fact: "How can this center be found?" "Can a figure have more than one turn symmetry?" "Could you define figures only in terms of their turn symmetries?" Some books define *point symmetry;* teachers can ask students how it is related to turn symmetry. Teachers can also suggest that students look in the phone book and see how many business logos they can find that have turn symmetry.

Glide Reflections

Principles and Standards for School Mathematics discusses only reflection, translation, and rotation transformations. However, another type of transformation must be considered to complete a survey of the structure of transformations—the *glide reflection*. This transformation is the

A glide reflection, like any reflection, reverses the orientation.

composition of a translation and a reflection, with the condition that the reflecting line for the reflection is parallel to the direction of the translation. An example is seen in figure 1.11. The original figure labeled A is translated to A', and then A' is reflected about a line parallel to $\overleftrightarrow{AA'}$ to obtain A''.

On the basis of the activities Slide Me Now and Fold Me! Flip Me! students should be able to find two parallel reflecting lines that determine the translation, as well as one reflecting line for the final reflection, making three reflections that will accomplish the glide reflection. The activity Gliding Along leads students through steps to find the reflecting lines. Among the things that they will discover is that a glide reflection, like any reflection, also reverses the orientation. In addition, the reflecting lines that determine the translation in a glide reflection are perpendicular to the reflecting line of the glide reflection.

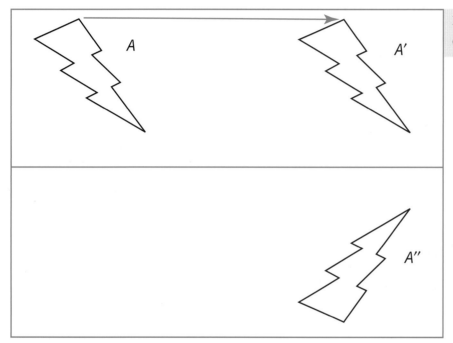

Fig. **1.11.**

Glide reflection

Gliding Along

Goals

- Understand properties of a glide reflection
- Construct a glide reflection as the composition of reflections in a pair of parallel lines and a line perpendicular to them
- Establish conjectures and proof by means of constructions

Materials and Equipment

- A copy of the activity pages for each student
- A Mira or an equivalent reflective device for each student

Discussion of the Activity

p. 85

From a mathematical point of view, a primary reason for including glide reflections in our discussion of isometries is that the set of all transformations described in this chapter form a group under the operation of composition. Students may begin to appreciate this fact when they discover that for any two given congruent figures in a plane, it takes no more than the composition of three reflections to map one figure to the other. The complete demonstration of this is beyond the scope of this book but not beyond the construction methods of all students. (For a description of the constructions necessary to argue this, see Dayoub and Lott 1977.)

An Interdisciplinary Application of Transformations

Olson's construction of the tangents enveloping a parabola is shown on the accompanying CD-ROM.

Olson (1975) describes how to construct tangents to a parabola by paper folding. The tangents "envelop" the parabola and can be used to find points on it. The same constructions can be accomplished using either a reflective tool such as a Mira or geometry software. The activity Into the Light with Transformations demonstrates the constructions using a TI-92 Plus calculator.

Into the Light with Transformations

Goals

- Use properties of a reflection in an application
- Construct a parabola using the notion of an envelope of lines
- Use data collection and quadratic regression to find the equation of a model
- Establish conjectures and proof by means of constructions

Materials and Equipment

- A copy of the activity pages for each student
- TI-92 or TI-92 Plus calculator with Cabri Geometry II software (or a computer with Geometer's Sketchpad)

p. 87

Discussion of the Activity

The activity will enhance students' understanding of the mathematics by using a variety of technological methods and representations to "prove" that the points obtained in different representations are actually on a particular parabola. The first of these methods "shows" with technology that all points P (see fig. 1.12a) are the same distance from the directrix and the focus and hence must be on the parabola. That is, students show that $PF = PB$.

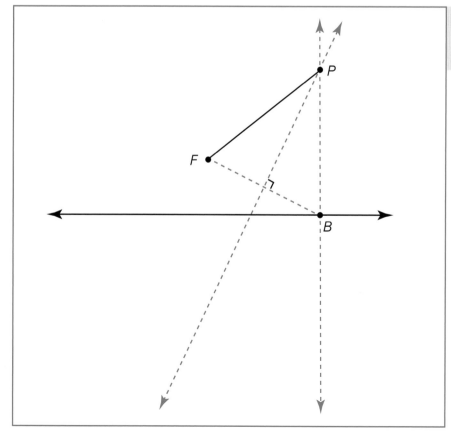

Fig. **1.12a.**

Different representations of a parabola

A second method uses △*FPB* (see fig. 1.12b) to generate conjectures about the relationship of point *P* to point *F* and line *DB*. Finally, a third method allows the coordinatization of the plane, collects the coordinates of the points found, and carries out a quadratic regression on this data.

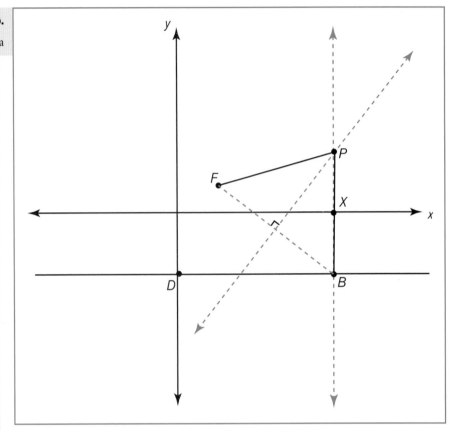

Teachers may recall from analytic geometry that a parabola with directrix parallel to the *x*-axis has an equation of the form $y = ax^2 + bx + c$. With its computer algebra system (CAS) and data-collection capabilities, a TI-92 Plus calculator can use a quadratic regression to find an equation that describes points on the parabola.

Transformations with Matrices

Principles and Standards for School Mathematics recommends that high school students learn to represent reflections, rotations, and translations with matrices and that they use them to explore the properties of the transformations. Furthermore, "students should understand that multiplying transformation matrices corresponds to composing the transformations represented" (p. 315). For example, in figure 1.13, △*ABC* with vertices located at (–5, 1), (–4, 7), and (–8, 5) is reflected in the line represented by the equation $y = x$.

In this section, we focus on the coordinates of a point and its image, suggested in *Principles and Standards*. The coordinates of the vertices of a figure can be represented in matrix form. For example, the vertices of △*ABC* in figure 1.13 can be represented as matrix *M* below:

$$M = \begin{bmatrix} -5 & -4 & -8 \\ 1 & 7 & 5 \end{bmatrix}.$$

In a reflection in the line $y = x$, the net result on the coordinates of the vertices of $\triangle ABC$ is that the x- and y-coordinates are reversed. For example, reflecting $(-5, 1)$ in the line yields $(1, -5)$. This type of reflection can be achieved by multiplying the matrix containing the coordinates of the vertices by the reflection matrix

$$r_{y=x} = \begin{bmatrix} 0 & 1 \\ 1 & 0 \end{bmatrix},$$

as seen in the following:

$$r_{y=x}M = \begin{bmatrix} 0 & 1 \\ 1 & 0 \end{bmatrix}\begin{bmatrix} -5 & -4 & -8 \\ 1 & 7 & 5 \end{bmatrix} = \begin{bmatrix} 1 & 7 & 5 \\ -5 & -4 & -8 \end{bmatrix}.$$

As *Principles and Standards* directs, transformations can be composed by multiplying matrices. However, if we consider a translation that shifts each x-coordinate 5 units to the right and leaves the y-coordinates unchanged, as in figure 1.14, we encounter problems in multiplying matrices. The dimensions of the matrices for the translation and for the vertices of the triangle do not allow the multiplication.

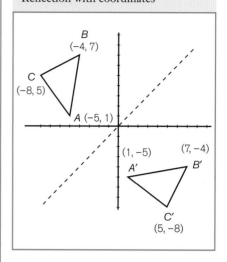

Fig. **1.13.**

Reflection with coordinates

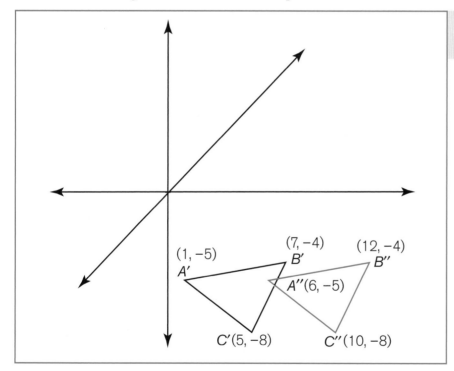

Fig. **1.14.**

Translation of $\triangle A'B'C'$ by $(5, 0)$

In order to achieve a translation that adds 5 to the x-coordinate and 0 to the y-coordinate of the vertices of the triangle, matrix addition—not multiplication—must be used. Because the coordinate matrix of the triangle has two rows and three columns, the translation matrix must also have these dimensions if the two matrices are to be added. A matrix that can be used to achieve the translation is the following:

$$T = \begin{bmatrix} 5 & 5 & 5 \\ 0 & 0 & 0 \end{bmatrix}.$$

The translation could then be accomplished as follows:

$$T + M = \begin{bmatrix} 5 & 5 & 5 \\ 0 & 0 & 0 \end{bmatrix} + \begin{bmatrix} 1 & 7 & 5 \\ -5 & -4 & -8 \end{bmatrix} = \begin{bmatrix} 6 & 12 & 10 \\ -5 & -4 & -8 \end{bmatrix}.$$

Now suppose that we would like to compose the reflection in the line with the translation just described. The matrix representing the translation was previously adjusted to have two rows and three columns, but in its simplest form it could be represented with two rows and one column:

$$\begin{bmatrix} 5 \\ 0 \end{bmatrix}.$$

(This is also the matrix for the vector used in the translation.) Neither of these representations allows us to multiply the translation matrix by the matrix representing the reflection:

$$\begin{bmatrix} 0 & 1 \\ 1 & 0 \end{bmatrix}.$$

In fact, the dimensions of the matrices prevent us from either adding or multiplying them to achieve the desired composition.

To overcome this difficulty, Pettofrezzo (1966) suggests rewriting all transformation matrices as 3×3 matrices. Doing so requires the use of a dummy row that has no effect on either the coordinates of the original or the reflection matrix but still allows us to achieve the translation.

Pettofrezzo's recommendation builds on the fact that transformations involving 2×2 matrices deal with functions known as homogeneous linear transformations of the plane, which take a point with coordinates (x, y) to a point with coordinates $(x', y') = (ax + by, cx + dy)$. This transformation could be represented generally as

$$\begin{bmatrix} x' \\ y' \end{bmatrix} = \begin{bmatrix} a & b \\ c & d \end{bmatrix} \begin{bmatrix} x \\ y \end{bmatrix} = \begin{bmatrix} ax + by \\ cx + dy \end{bmatrix}.$$

To consider translations of the plane, we must deal with nonhomogeneous linear transformations of the plane that take a point with coordinates (x, y) to a point with coordinates $(x', y') = (ax + by + c, dx + ey + f)$. A transformation of this type could be represented with matrices generally as

$$\begin{bmatrix} x' \\ y' \\ 1 \end{bmatrix} = \begin{bmatrix} a & b & c \\ d & e & f \\ 0 & 0 & 1 \end{bmatrix} \begin{bmatrix} x \\ y \\ 1 \end{bmatrix} = \begin{bmatrix} ax + by + c \\ dx + ey + f \\ 1 \end{bmatrix}.$$

Sibley (1998) describes the process differently (but equivalently):

> Mathematicians have devised a simple way around this problem by using as their model the plane z in $\Re^3$. Clearly, it has the same geometric properties as $\Re^2$, which, in effect, is the plane $z = 0$. However, $z = 1$ has the key algebraic advantage that all its points $(x, y, 1)$, including the new "origin"

If we rewrite transformation matrices as 3 × 3 matrices, we can multiply them to determine the composition of transformations of the plane.

(0, 0, 1), can be moved by 3×3 matrices. The third coordinate of $(x, y, 1)$ does not really "do" anything. (P. 146–47)

Consider how the translation in figure 1.14 could be accomplished with a matrix. Remember that we want a point with coordinates (x, y) to be translated and to have an image with coordinates $(x', y') = (x + 5, y + 0)$. This could be accomplished with a 3×3 matrix, as seen below:

$$\begin{bmatrix} x' \\ y' \\ 1 \end{bmatrix} = \begin{bmatrix} 1 & 0 & 5 \\ 0 & 1 & 0 \\ 0 & 0 & 1 \end{bmatrix}\begin{bmatrix} x \\ y \\ 1 \end{bmatrix} = \begin{bmatrix} 1x + 0y + 5 \\ 0x + 1y + 0 \\ 1 \end{bmatrix} = \begin{bmatrix} x + 5 \\ y \\ 1 \end{bmatrix}.$$

With this type of notation, the reflection in the line $y = x$ could be represented by the matrix

$$\begin{bmatrix} 0 & 1 & 0 \\ 1 & 0 & 0 \\ 0 & 0 & 1 \end{bmatrix},$$

with the annexation of an additional row and column to the previously used matrix,

$$\begin{bmatrix} 0 & 1 \\ 1 & 0 \end{bmatrix}.$$

Thus, the composition of the reflection in the line, followed by a translation that increases each x-coordinate by 5 and leaves the y-coordinate unchanged, can be determined by multiplying the matrices, as follows:

$$\begin{bmatrix} 1 & 0 & 5 \\ 0 & 1 & 0 \\ 0 & 0 & 1 \end{bmatrix}\begin{bmatrix} 0 & 1 & 0 \\ 1 & 0 & 0 \\ 0 & 0 & 1 \end{bmatrix}\begin{bmatrix} -5 & -4 & -8 \\ 1 & 7 & 5 \\ 1 & 1 & 1 \end{bmatrix} = \begin{bmatrix} 6 & 12 & 10 \\ -5 & -4 & -8 \\ 1 & 1 & 1 \end{bmatrix}.$$

Note that, using matrices, the final coordinates of $\triangle A''B''C''$ are the same as the final coordinates of the image triangle in figure 1.14. An investigation of matrix representations of transformations is seen in the activity Transforming with Matrices.

Transforming with Matrices

Goals

- Use matrices to represent transformations
- Find coordinates of images of a figure using matrices
- Determine *invariants*, or unchanging properties, of transformations

Materials and Equipment

p. 95

- A copy of the activity pages for each student
- Graph paper for each student
- A graphing calculator with matrix operations for each pair of students (the activity can also be completed with hand computations)

Discussion of the Activity

In Transforming with Matrices, students explore the multiplication possibilities of a 3×3 coordinate matrix with a dummy last row of ones. They draw the triangle that we have just discussed, with coordinates $(-5, 1)$, $(-4, 7)$, and $(-8, 5)$. They reflect it in the line $y = x$ and translate it by $(5, 0)$ (see figs 1.13 and 1.14). They then repeat the transformations, this time using matrix multiplication, and they verify the coordinates that they obtain for the composition by comparing them with their earlier results.

Students use matrices to think more generally about coordinates under a reflection, a translation, and a rotation as they answer Discussion and Extension questions.

Conclusion

By including matrices, a study of transformations achieves an integration of traditional geometry, analytic geometry, and algebra. Understanding these connections is important in a number of fields, including animation in the areas of computer science, film making, and robotics. Chapter 2 expands some of these connections in its examination of the art of map making.

NAVIGATIONS SERIES

GRADES 9–12

NAVIGATING *through* GEOMETRY

Chapter 2
The Geometry of Location and Map Making

Chapter 1 prepared teachers and students to use transformations as mathematical tools for exploring other geometric topics. *Principles and Standards for School Mathematics* (NCTM 2000) suggests that students "specify locations and describe spatial relationships using coordinate geometry and other representational systems" and be able to "use Cartesian coordinates and other coordinate systems, such as navigational, polar, or spherical systems, to analyze geometric situations" (p. 308). Applying transformation geometry to reflect these Standards leads from coordinate systems to map making. Though many of the transformations used in map making extend beyond the mathematics typically encountered by high school students, the properties of projections involved in map making are well within their reach.

Map making, or cartography, is an old science, dating to about 2300 BC. In today's world, geometry is basic to both map making and to the Global Positioning System (GPS). Both incorporate ideas of *earth measure*, which is the literal meaning of *geometry*. In this chapter, we look at various representations of points, including points on the earth, using a variety of coordinate systems. In addition to locating points using the GPS, we consider geometric transformations used for map making. Although these transformations are unlike those discussed in chapter 1, they are also based on the mathematical concept of a function. For most of the transformations presented in this chapter, the domain is not a plane, as in chapter 1, but either a sphere or a hemisphere, and the range is typically a flat surface—either a plane or a part of a plane.

Interested readers will find most mathematical transformations used by today's cartographers described in "Map Projections—A Working Manual" (Snyder 1987).

A time line of
milestones in
the history of
map making can be seen on
the CD-ROM
accompanying this book.

Coordinate Systems in a Plane

The traditional coordinate system in a plane is the rectangular Cartesian system named for René Descartes and based on two perpendicular axes, as depicted in figure 2.1a. Equally useful is the polar coordinate system, shown in figure 2.1b, which is based on the distance measured from a point called a pole and a polar angle measured from a diameter called the polar axis. Both of these systems are very valuable for describing locations in mathematics as well as in the physical world.

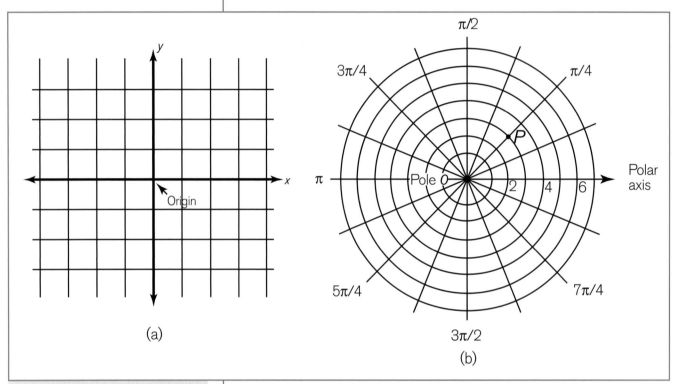

(a)

(b)

Fig. 2.1.

Rectangular and polar coordinate systems

A good geometric or algebraic exercise for students is to determine the transformation that allows conversions between the two coordinate systems in figure 2.1. Superimposing the two axes from figure 2.1 allows students to discover that if a point is represented by $P(x, y)$ in the rectangular coordinate system and $P(r, \theta)$ in the polar coordinate system, then

$$r = \sqrt{x^2 + y^2}$$

and

$$\theta \text{ is the angle whose tangent is } \frac{y}{x} \left(\text{or } \theta = \arctan\left(\frac{y}{x}\right)\right).$$

Thus, a point with coordinates (r, θ) in the polar system can be found from the corresponding point with coordinates (x, y) in the rectangular system by means of the following equations:

$$r = \sqrt{x^2 + y^2}$$

$$\theta = \arctan\left(\frac{y}{x}\right).$$

Latitude and Longitude

On a globe, a grid system of latitudes and longitudes can be used to identify locations on Earth, much as the *x* and *y* grid lines of the Cartesian coordinate system identify locations on a plane. Longitudes are *great circles*—that is, circles whose centers are at the center of the earth—running north and south along the earth, as seen in figure 2.2a. Longitudinal lines are sometimes known as *meridians.* In former times, the meridian or longitude that was chosen as the starting line of the measuring system usually depended on which country was doing the measuring. This practice changed in 1884, when the meridian through Greenwich, England, at the site of the Royal Observatory, was designated 0° longitude and was declared the *prime meridian* for the entire earth. (This meridian is also called the Greenwich Meridian.)

Latitude runs east and west on the earth and measures distance north and south from the equator, as seen in figure 2.2b. Latitudinal "lines" form circles that have smaller circumferences as they get closer to the poles, and the equator is the only latitude that is a great circle. Geographers sometimes talk about "lines" of latitude and longitude on the globe, but *line* is often a misnomer in this context. Although longitudes are great circles, which are considered lines in spherical geometry, latitudes, except for the equator, are not great circles. A latitude is a circle on the sphere, and the set of latitudes is comparable to concentric circles in the plane of Euclidean geometry: they do not intersect, yet they are not parallel, since they are not lines.

The equator is designated as 0° latitude, and the North and South Poles are 90° north and 90° south, respectively. Just as the *x*- and *y*-axes are used as references in the rectangular coordinate system, so 0° latitude (the equator) and 0° longitude (the prime meridian) serve as references in the latitude-longitude grid system for locating points on the earth. As seen in drawings that use the rectangular coordinate system, where usually only the axes are marked and a scale is shown, maps and globes generally show only a few latitudes and longitudes. A system of latitude and longitude is seen on the globe in figure 2.2c.

The Web site of the Royal Observatory in Greenwich, England (www.rog.nmm.ac.uk/), provides useful information. Hammond World Atlas Corporation has another valuable site (www.hammondmap.com /latlong.html), which offers a primer on latitude and longitude. Clicking on "Projection Animations" presents various map projections, and the Hammond Projection Applet includes animated maps that change from one projection to another, with explanations provided.

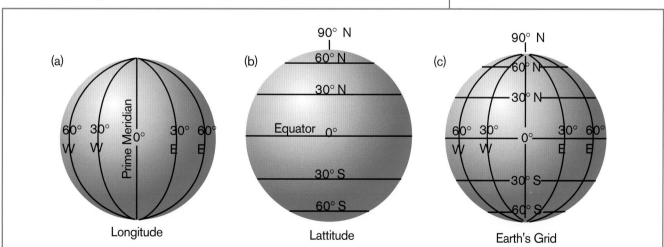

(a) Longitude **(b)** Lattitude **(c)** Earth's Grid

Fig. **2.2.**

Latitudes, longitudes, and a grid system on a globe

How does the latitude-longitude system locate a point on the globe? Degrees of latitude and longitude are measured as *central angles*, each of which has its vertex at the center of the earth. This is illustrated in figure 2.3. Degrees of latitude are measured with the initial side of the angle on the plane of the equator and the terminal side of the angle north or south of the equatorial plane. Degrees of longitude are measured on the plane of the equator, with the initial side of the angle on the plane of the prime meridian. The great circles of longitude intersect at the poles. The length of one degree of longitude is greatest at the equator (where it is approximately 69 miles) and contracts as the meridians approach the poles. The distances separating circles of latitude are fairly constant (approximately 69 miles) per degree. Thus, knowing that Rocky Mountain National Park is "about 40° north" gives a general idea of where it is located in relation to the equator. The city of Minneapolis, Minnesota, which is located at 44° 53′ N latitude and 93° 13′ W longitude, is approximately halfway between the equator and the North Pole and one quarter of the way around the earth from the prime meridian heading west.

Fig. 2.3.

Angles with vertices at the center of the earth

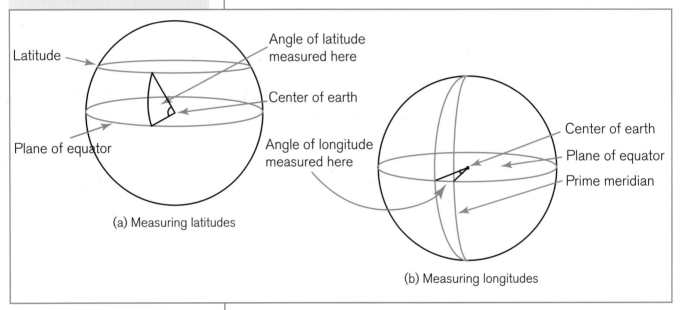

(a) Measuring latitudes

(b) Measuring longitudes

The story of John Harrison's forty-year struggle to solve the longitude problem is recounted in Longitude: The True Story of a Lone Genius Who Solved the Greatest Scientific Problem of His Time *(Sobel 1995). Some of Harrison's timekeeping devices can be viewed at www.rog.nmm .ac.uk/museum/harrison /index.html.*

As suggested by the measurements for Minneapolis, degrees of latitude and longitude are subdivided for greater precision into fractional parts called *minutes*. There are 60 minutes (′) in each degree of latitude or longitude. The minutes are also subdivided, into *seconds*, with 60 seconds (″) per minute. In practice, however, seconds of longitude and latitude are typically reported as decimal fractions of a minute and are shown to the nearest thousandth. For example, Alberta Falls in Rocky Mountain National Park is at 40° 18.361′ north and 105° 38.377′ west.

How to determine longitude at sea was a problem that achieved considerable fame, challenging navigators for centuries. In 1675, King Charles II of England founded the Royal Observatory to solve the problem, and in 1714, Parliament offered a prize of £20,000 to the person who could devise a method to measure longitude at sea to within one-half of a degree (two minutes of time). A clockmaker named John Harrison worked on the problem for forty years before solving it successfully.

Navigating through Geometry in Grades 9–12

Global Positioning System

We have seen how the lines comprising the *x*-axis and the *y*-axis are used to locate points in a rectangular coordinate system, how a circle and a ray are used to locate points in the polar coordinate system, and how the intersecting circles of latitudes and longitudes are used to locate points on the surface of a globe. A new position-locating system has evolved in recent years—one that is based on the intersection of spheres. The Global Positioning System (GPS) is a satellite-based navigation system that locates positions on Earth at any time of day in any weather. GPS is short for NAVSTAR GPS, which stands for NAVigation Satellite Timing And Ranging Global Positioning System. The concept behind the GPS was born in 1973, when branches of the U.S. military pooled resources to create a satellite-based navigation system. After years of feasibility testing, the first GPS satellite was launched in 1978. The implications and applications of the mathematics—and in particular the geometry—of the GPS are just now beginning to be realized. Studying the GPS helps students in grades 9–12 reach the goal of possessing "some familiarity with ... systems used in navigation" (NCTM 2000, p. 314).

Parts of the GPS

The GPS has three parts: the space segment, the control segment, and the user segment. Each plays an important role in determining position. However, the space and control segments are largely invisible to the user. The space segment consists of twenty-four satellites. Twenty-one are usually active, and three serve as spares that can be activated if others malfunction. The satellites are in six orbits of four satellites each, positioned in such a way as to make at least five satellites "visible" to a GPS receiver at any time of day from any place on Earth. The satellites make two complete orbits of the earth in just under twenty-four hours. The satellites' orbits are approximately 11,000 miles above the earth, a height that makes them predictable, controllable, and impervious to the earth's atmosphere. Each satellite has an atomic clock that is accurate to about one second in 70,000 years. This accuracy is at the heart of the process of determining a location on Earth.

The control segment of the GPS consists of a master control station that constantly monitors, tracks, and manages the satellites. Its duties include predicting satellite orbits, keeping the satellites in the orbits, activating spare satellites, and monitoring all satellite transmissions.

The user segment of the GPS captures the satellite information necessary for determining location, velocity, or time. Originally, GPS users were involved primarily in military operations, but civilian users now far outnumber military users. Central to the user segment is a GPS receiver. Some receivers are built into applications, such as an automobile's navigational system or a farm tractor's onboard computer; others are handheld models used in activities such as hunting, fishing, or hiking. New applications of the GPS are continually being discovered. The activity Delivering Packages can serve to acquaint students with the system.

For a useful overview of the GPS on the Web, go to www.colorado.edu/geography /gcraft/notes/gps/gps_f.html or to www.trimble.com.

A few GPS applications are listed on the CD-ROM accompanying this book.

Delivering Packages

Goals

- To use coordinates of latitude and longitude with locations on a map
- To determine the most efficient travel route for a package delivery driver

Materials and Equipment

- One copy of the activity pages for each student
- One overhead pen and transparency for each group of students

Discussion of the Activity

p. 97

Though a GPS is not required for Delivering Packages, the thinking that the activity involves is related to the mathematics that a GPS uses to locate points and routes. To introduce students to the uses of the GPS, the activity directs them to use a detailed map to try to find the "best"—that is, the most efficient—route for a delivery person to take in dropping off seven packages in the Anoka, Minnesota, area. Teachers should be prepared for some students to mistake the map's grid lines for roads and to propose routes that would have the driver traveling on these lines.

After students have worked individually, they then meet in groups to compare their ideas and agree on the "best" route. They trace this route on an overhead transparency. Teachers can put all the transparencies on the overhead projector and discuss their differences.

How the GPS Works

To compute distance from a satellite, a GPS receiver enters the travel time of signals sent from the satellite into the familiar formula "distance equals rate times time." Because the signals are sent from the satellite at the speed of light (approximately 186,000 miles per second), all a receiver needs to do to determine its distance from the satellite is to compute the difference between the time when the signal was sent and the time when it was received and multiply that difference by the speed of light. (See Dahl 1993 and House 1997.)

To understand the process by which the GPS determines the user's position on the earth, assume that a GPS user somewhere in the universe is 11,000 miles from a satellite, as determined from the signals sent by that satellite. This means that the user is somewhere on a sphere centered at that particular satellite, with a radius of 11,000 miles, as seen in figure 2.4. (Note that this sphere is not the earth but the set of points 11,000 miles from the satellite.)

Also assume that the user is 11,500 miles from a second satellite. This means the user is on a sphere centered at the second satellite, with a radius of 11,500 miles. Because the user is on the intersection of two spheres, the user must be on a circle, as shown in figure 2.5.

See House (1997) on the CD-ROM for additional information on the mathematics of the GPS.

Fig. 2.4.

A GPS user 11,000 miles from one satellite

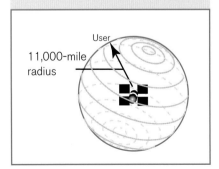

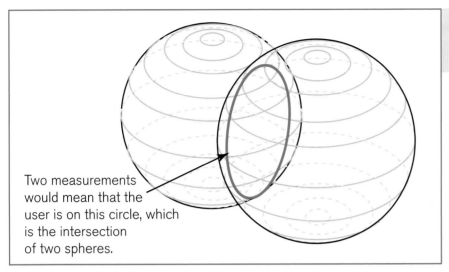

Fig. **2.5.**

A GPS user 11,500 miles from a second satellite

Two measurements would mean that the user is on this circle, which is the intersection of two spheres.

Assume now that the GPS user is also 12,000 miles from a third satellite. Then the user must be at one of the two points that are the intersection of the circle (from fig. 2.5) with the sphere of radius 12,000 miles, as seen in figure 2.6.

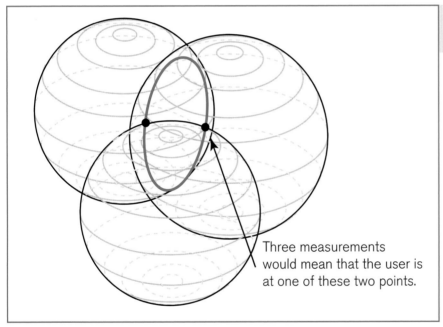

Fig. **2.6.**

The GPS user at one of two points

Three measurements would mean that the user is at one of these two points.

Of the two possible points, one is likely to be wildly improbable, suggesting that the user is deep inside the earth or far out in space. The receiver rejects the unrealistic possibility and shows the other as the position of the user. Most receivers eliminate one of the two possibilities by using information from a fourth satellite or by knowing in advance the general location of the user.

GPS Signal Code

A GPS receiver uses a *pseudorandom* code in calculating the difference between the time when a signal was sent and the time when it was received. Each satellite sends signals to Earth in a binary code (made up only of 0s and 1s), as in figure 2.7. These signals appear at first to be

An excellent source for information about how a GSP works is "Global Positioning System: The Mathematics of GPS Receivers" (Thompson 1998).

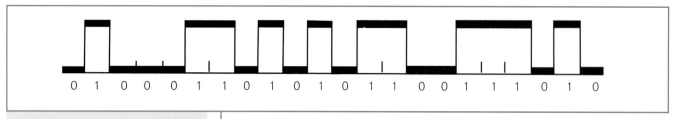

| 0 | 1 | 0 | 0 | 0 | 1 | 1 | 0 | 1 | 0 | 1 | 0 | 1 | 1 | 0 | 0 | 1 | 1 | 1 | 0 | 1 | 0 |

Fig. 2.7.

Sample signals sent in pseudorandom code

random but on closer inspection are seen to form precise patterns—hence the name pseudorandom. The signals sent by the satellites need to be complicated so that the receiver does not accidentally synchronize with some other, stray signal. Also, each satellite has a unique set of signals so that the receiver will not pick up another satellite's signal by mistake. The signals include information about the exact time when they left the satellite, as marked by its atomic clock.

The satellite sends a radio signal in pseudorandom code, and the receiver generates an internal code that is an exact duplicate of the satellite's code. However, because it takes time for the satellite's signal to reach the receiver, the two signals do not precisely match. This mismatch reflects the time delay from the satellite to the receiver. The receiver compares the satellite's signal to its own, and by analyzing the mismatch, determines how much time the satellite's signal took to reach the receiver. This measured time can then be used in the relationship "distance equals rate times time" to determine the satellite's distance from the receiver. Figure 2.8 shows one such shift and the time difference it reflects. Although the clock in the GPS receiver is not an atomic clock, it is constantly synchronized with the satellite's clock, so it gives a very accurate measure.

Fig. 2.8.

Difference in time between satellite- and receiver-generated signals

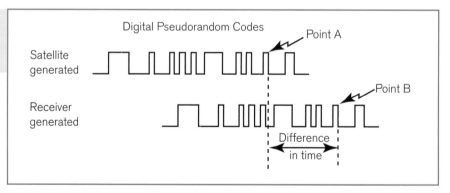

Airlines use the GPS to find *way points*—that is, points along the route of a plane as it travels from one airport to another. On some planes, passengers can track the plane's route on video screens. Time in those displays is given as Greenwich Mean Time (GMT), which is mean solar time of the meridian at Greenwich, England. GMT is used as the basis for standard time throughout most of the world. The activity Where Are We Now? allows students to track a plane on a trip from the Minneapolis–St. Paul International Airport to the San Diego International Airport.

Where Are We Now?

Goals

- To use latitude and longitude coordinates in locating way points—
  places between the main points on a travel route—on a map of the
  United States

Materials and Equipment

- A copy of the activity pages for each student

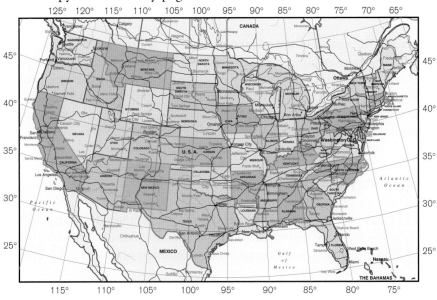

p. 99

Discussion of the Activity

Where Are We Now? asks students to use the latitudinal and longitudinal coordinates of Minneapolis–St. Paul International Airport and San
Diego International Airport, as well as the coordinates of four different
way points on a plane's flight between the two terminals, to decide which
direction the plane is traveling. In addition, Discussion and Extension
questions invite the students to use the times at which the coordinates of
the way points were recorded to calculate approximate times for the
plane's takeoff, landing, and total flight. To make these calculations, students need to assume that the plane's speed is constant and that its flight
path is linear. These assumptions allow students to manipulate and think
about the data at hand. Working with them, the students can set the
ratio of a known distance to a known time equal to the ratio of another
known distance to an unknown time that they want to determine.

The flight from Minneapolis to San Diego travels through only
twelve degrees of latitude, and the latitudes of two of the way points
differ from each other by less than one degree. Thus, it seems reasonable for students to use this linear method. However, some teachers
may want to give students a more accurate and detailed appreciation for
the spherical geometry and curvature of the earth that are involved in
the activity's scenario. At Minneapolis (approximately 45° N)—halfway
between the equator and the North Pole—the distance that corre-

sponds to one degree of longitude is less than it is at San Diego (approximately 32° N). Likewise, at San Diego, the distance that corresponds to one degree of longitude is less than the approximately sixty-nine miles per degree at the equator. Becoming aware of these differences will help students understand the imprecision of linear proportional reasoning.

A formula for computing an approximate distance between two points, given their latitudes and longitudes, provides an interesting extension for trigonometry students. Consider city A with latitude a_1 and longitude b_1, and city B with latitude a_2 and longitude b_2. The distance between the cities is given by the following:

$$\{[\arccos(\cos a_1 \cdot \cos b_1 \cdot \cos a_2 \cdot \cos b_2 + \cos a_1 \cdot \sin b_1 \cdot \cos a_2 \cdot \sin b_2 + \sin a_1 \cdot \sin a_2)]/360\} \cdot 2\pi r,$$

where r is the radius of the earth (approximately 3693 miles at the equator). The calculation will yield a fraction of the circumference of the earth that, when multiplied by $2\pi r$ (the earth's circumference), will give a distance on the earth. Substituting the coordinates of Minneapolis and San Diego into the formula yields a distance of 1531 miles, compared to the accepted distance of 1526 "as the crow flies."

Learning with Maps

Just as the GPS is changing methods of determining location in today's world, so the mathematics of map making has changed over time. From the days of the simple maps identified on the time line on the CD-ROM, transformations for map making have changed, changing the cartographer's job, as well. To begin to think about map making, consider that maps give us information about directions, population trends, street layouts, and land masses, among other topics. With all our daily use of maps, it is sometimes surprising to think that there is a major problem in making a map of the earth. The earth is an oblate spheroid. Flattened at the poles, it is almost but not quite a sphere. The circumference of the earth at the equator (approximately 24,901 miles) is about 42 miles greater than the circumference of a great circle going through the poles (approximately 24,859 miles). A globe provides the best model of the earth's surface, but it is not always the most realistic or practical one, since we usually want to view areas much smaller than the entire earth. Most maps, then, attempt to represent the earth as accurately as possible on a flat surface.

Since the earth is approximately spherical, it is impossible to portray its features accurately on a flat surface and still give true representations of shapes, areas, and scales. The task of mapping local areas is less difficult. That is, we can draw a fairly realistic representation of a kitchen or even a town without worrying too much about the curvature of the earth. The same cannot be said of drawing a map of the entire northern hemisphere. From a mathematical standpoint, in small areas—locally—the earth is Euclidean, whereas in large areas—globally—it is not.

Projections in Map Making

A projection is one way to depict the surface of the earth on a plane.

The accompanying CD-ROM presents a very different coordinate system with applications in mathematics—a trilateral coordinate system.

A Rice University Web site at www.ruf.rice.edu /~feegi/site_map.html offers insights into map making. Links under "Maps and Charts" take users to sites on latitude, longitude, map making, astronomy, and other topics, including "Math in Maps," where clicking on "Cartography Document" gives access to a PDF file on latitude, longitude, and maps, with information on the mathematics behind some of the distortions involved in cartography.

It uses a transformation that essentially maps a sphere onto a plane. Many different projections are used to make a variety of maps. The type of projection that a mapmaker uses depends on the properties of the earth (considered as a sphere) that he or she wants to preserve on the map (considered as a plane). Some projections preserve shape; others preserve area or distance or direction or angle measure or combinations of these. However, no single projection preserves all properties of the sphere when transforming it to the plane. In the material that follows, we examine some common projections and consider the transformations used to create them.

Three main types of projections—*azimuthal, cylindrical,* and *conical*—are used by cartographers to map the earth to the plane. Azimuth refers to the arc of the horizon as measured from a pole, and azimuthal projections are those in which the shortest distance between any point and the center is a straight line representing a great circle through a pole. In the activity Intuitive Cartography, students begin to think about preserved properties in different azimuthal projections, and they identify some scales that remain "true" and others that distort what they represent.

Three main types of projections—azimuthal, cylindrical, and conical—are used by cartographers to map the earth to the plane.

Intuitive Cartography

Goals

- To determine properties preserved in map making
- To simulate a gnomonic azimuthal projection (Task *A*)
- To simulate a stereographic azimuthal projection (Task *B*)

Materials and Equipment

p. 101

- A copy of the activity pages for each student

For Task *A*—

- A round, transparent (plastic) hemispheric object for each pair of students. (The hemispheric transparency for the Lenart Sphere is one such object. Also, clear plastic spheres that come apart into hemispheres are readily available in craft stores.)
- A sheet of unlined paper for each pair of students (the size will depend on the size of the hemispheric object—the larger the object, the larger the paper)
- Overhead markers (at least two colors) for each pair of students
- A flashlight for each pair of students

For Task *B*—

- A lamp (without a shade or a shade harp) for each group of students
- A clear spherical light bulb (of the type commonly found in bathroom fixtures) for each group of students
- Marking pens for white boards or colored chalk for chalkboards (at least two colors) for each group of students
- Marking pens (at least two colors per group) for drawing on the light bulbs
- Depending on the electrical outlets in the classroom, an extension cord for each group of students

Note: Task *B* involves materials that some teachers may be unable to obtain in sufficient quantities for students to use in small groups. In addition, some classrooms may lack sufficient numbers of electrical outlets or have fire and safety regulations that prohibit the use of multiple extension cords. If necessary, teachers can conduct Task *B* as a whole-class activity, with just one lamp base and one light bulb.

Discussion of the Activity

Intuitive Cartography emphasizes visualization and spatial reasoning. *Principles and Standards* states that "students should develop visualization skills through hands-on experiences with a variety of geometric objects" (p. 43). By drawing shapes on plastic spheres and light bulbs and projecting them onto paper or a whiteboard or chalkboard, students physically transform a sphere into a projection on a plane in much the same way that cartographers create maps.

The activity consists of two self-contained tasks, both of which involve placing a "globe" tangent to a plane and projecting the globe onto it. Task *A* simulates a *gnomonic* projection, and Task *B*, a *stereographic* one. Gnomonic and stereographic projections are two of several types of azimuthal projections. Gnomonic projections, which take their name from *gnomon*, an early instrument for determining a position's latitude by the length of its noontime shadow, show points on a hemisphere as projected from the center of the sphere onto a plane tangent to the hemisphere at a pole (see fig. 2.9a). Stereographic projections also project points onto a plane tangent at a pole, but this time the points are projected from the sphere's opposite pole instead of from its center (see fig. 2.9b).

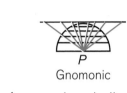

Gnomonic

A gnomonic projection of part of a sphere onto a tangent plane from point *P* (the center of the sphere)

(a)

Stereographic

A stereographic projection of part of a sphere onto a tangent plane from point *Q* (the pole opposite to the pole tangent to the plane)

(b)

Fig. **2.9.**

Azimuthal projections

Task *A* uses a clear plastic hemisphere to represent half of a globe in a simulation of a gnomonic projection. Students use markers to draw latitudes and longitudes on the hemisphere and hold a flashlight near its center to represent the point from which the projection is made. Task *B* simulates a stereographic azimuthal projection, using a spherical light bulb to represent a globe with a projection point at a pole. Unfortunately, the simulation is only approximate because the filaments of most bulbs are closer to the center of the bulbs than to a "pole." Consequently, the projection that results will resemble something between a stereographic projection and a gnomonic one.

However, both tasks permit students to estimate the properties that are preserved in the two types of projections. They may be tripped by the activity question about whether the projections preserve parallelism, since a globe has no parallel lines. This is an important lesson for the students to learn. They may suggest railroad tracks as sets of parallel lines. Considered from a local vantage point, the tracks appear to be parallel lines, but they do not qualify as lines when considered as figures on a globe. The following question is a good one to ask students in connection with the activity Intuitive Cartography: "What shapes get mapped as straight lines in the projections that you explored?" In both gnomonic and stereographic projections, all great circles through the pole are mapped as straight lines. This property is especially useful in navigation for planning airplane routes. It explains why airplanes that fly from Boston to Tokyo fly near the North Pole.

In both gnomonic and stereographic projections, all great circles through the pole are mapped as straight lines.

Fig. **2.10.**

Cylindrical projection

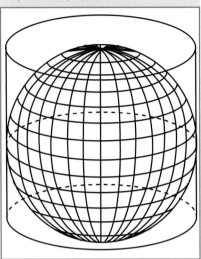

Cylindrical Projections

Many centuries-old cartographic techniques are still used today, as noted by Snyder (1987). One such technique was developed by the sixteenth-century Flemish cartographer Gerardus Mercator. Mercator's projection is a cylindrical one. He imagined a cylinder wrapped around the earth (or globe), as shown in figure 2.10, and projected the earth's sphere from its center onto the cylinder. In Mercator projections, the cylinder usually touches the earth at the equator.

The activity Projecting to a Cylinder allows students to undertake a hands-on investigation of the properties of a projection of the Mercator type. (This activity is adapted from a module, "What Shape Is Your World?" developed by the Montana Council of Teachers of Mathematics [1998]).

Projecting on a Cylinder

Goals

- To determine properties preserved in map making
- To simulate a cylindrical projection

Materials and Equipment

- A copy of the activity pages for each student
- A styrofoam hemisphere for each group of students (*Note:* Styrofoam spheres from craft stores can be halved easily.)
- Approximately five wooden skewers for each group of students
- Unlined white paper (large enough to form into a cylinder around the base of the hemisphere)
- A ruler for each group of students
- A protractor for each group of students
- Washable markers for each group of students (*Note:* If the hemispheres will not be reused, ballpoint pens will be adequate.)

p. 105

Discussion of the Activity

A cylindrical projection yields a map with straight lines representing longitude; latitudes are projected as circles on the cylinder, and when the cylinder is cut along a longitude and flattened, the latitudes appear as lines perpendicular to the longitudes. The activity helps students discover that the scale is true at the equator but that distortion increases at higher latitudes in both directions. Sailors often use cylindrical projections because they can plot a straight line segment between two points that are not very far apart and can continue on the path of the line segment without making constant corrections in their course.

Teachers who wish to pursue the study of coordinate systems can ask students how coordinates could be determined in a cylindrical projection. Some students may suggest using polar coordinates for a plane that is perpendicular to the cylinder, with the pole of the system at the center of a circular cross section of the cylinder. Then to locate a point P of the cylinder, they would need only to construct a perpendicular from P to the circular cross section, find the polar coordinates of the foot of the perpendicular, and determine the height of point P from the foot. The coordinates of point P are the triple consisting of the polar coordinates and the determined height.

Conical Projections

Another common projection used to create maps is a conical projection, which supposes that a cone of paper is placed over the globe, as shown in fig. 2.11, with the vertex of the cone on the line through the polar axis. The paper touches the globe along one latitude. The latitude-longitude grid is projected onto the cone from the center of the

Fig. **2.11.**

Conical projection

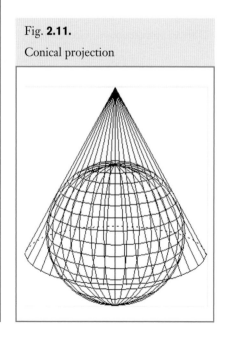

Fig. **2.12.**

Polyconic projections

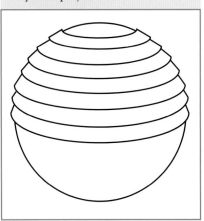

sphere. When the cone is opened flat, latitudes then appear as concentric arcs, and longitudes appear as straight lines radiating from the vertex of the cone. In this type of projection, the scale is true at the points of contact but distorted elsewhere. To overcome this limitation and to increase the accuracy of conical projection maps, cartographers use *polyconic projections*, in which they use narrow strips of different cones for the projections and then collect the strips together to make a single map (see fig. 2.12). Conical projections are often used for regional or country mappings rather than for views of entire hemispheres.

Regardless of the type of projection used to create a map, properties of the landscape may or may not be preserved. In this chapter, we have presented activities that allow students to determine intuitively some of those that are preserved. One example of a dramatic difference in preserved properties of stereographic and cylindrical projections is seen in figure 2.13. Both figure 2.13a and 2.13b show Greenland. Teachers can ask students to consider the differences in the two representations of the same land mass.

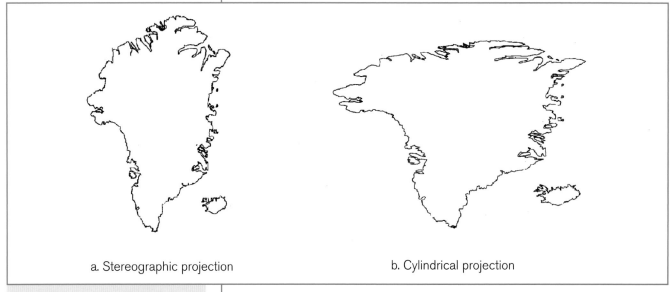

a. Stereographic projection

b. Cylindrical projection

Fig. **2.13.**

Different projections of Greenland

Conclusion

As techniques for locating points and map making have changed over the years, so must school mathematics. As more applications of the GPS are found, some see the system as a forerunner to some future navigation system. If a new system does indeed become a reality, mathematics will undoubtedly be at its center, and there will certainly be implications for school geometry. In the 1950s, most spherical geometry was removed from the secondary school curriculum. A knowledge of three-dimensional geometry and in particular spherical geometry now seems necessary for students to begin to understand the basics of the geometry behind the point-locating and map-making applications of today and those that may be developed in the future.

Principles and Standards points out that students should know about the geometry of navigation tools. Clearly, today's students need to understand more than the sextant and traditional maps. This chapter has presented a combination of coordinate systems (with some trigonometry), transformations, and three-dimensional geometry. The topics covered here perhaps give a glimpse of the future of geometry, taught and used not in isolation but integrated throughout mathematics.

In chapter 3, we equip students and teachers to investigate yet other developments in geometry. By focusing on similarity and congruence through transformations, we position readers to take a close look at the geometry of fractals in chapter 4.

NAVIGATING *through* GEOMETRY

Chapter 3
Multiple Dimensions of Similarity

We ground the idea of similarity in real-world examples that we use to explore relationships before examining the mathematics that is embedded in them.

In chapter 1, we laid the groundwork for using geometric transformations to unify the study of geometry, as *Principles and Standards for School Mathematics* (NCTM 2000) recommends. We built on this foundation in chapter 2, examining functions underlying geometric transformations that are used to convert from one coordinate system to another and that form the basis of map making. In chapter 3, we now consider geometric transformations underlying the traditional geometric notion of similarity. In the process, we will be setting the stage for expanding the idea of similarity to fractals in chapter 4.

As is appropriate in introducing many mathematical concepts to students, we ground the idea of similarity in real-world examples that we use to explore relationships before examining the mathematics that is embedded in them. A set of mixing bowls like that pictured in figure 3.1a, where each bowl has the same shape as any other but is a different size, can serve as one such example. Toy cars and trains that are replicas of full-size models, like the train shown in figure 3.1b, can serve as other examples. Consider also that architects, builders, scientists, sculptors, and engineers represent both the huge and the infinitesimal through scale models that they can manipulate and understand as they turn ideas and plans into reality. For instance, the artist Gutzon Borglum, sculptor of the Mount Rushmore National Memorial, made extensive use of scale models for his famous sculptures of the four U.S. presidents, as illustrated in figure 3.2a. The artists who are currently completing the mountain tribute to Crazy Horse, begun by Korczak Ziolkowski, are also using scale models, as shown in figure 3.2b.

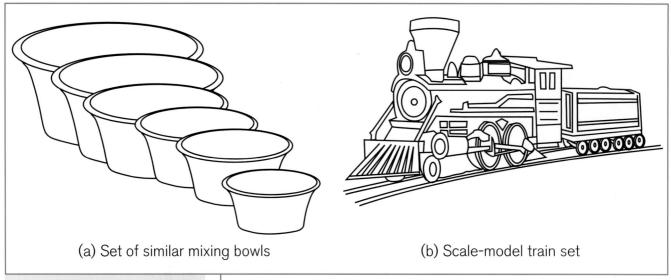

(a) Set of similar mixing bowls   (b) Scale-model train set

Fig. **3.1.**

Commonplace similar objects in real life

In these and countless other situations, we find a common mathematical characteristic: *similarity*. In mathematics, similar objects maintain identical shape even though they vary in size. To examine similarity, we continue to look at transformations, focusing now on those that preserve shape. We then examine two applications that include data-collection activities in which similarity plays an important problem-solving role.

Similarity and Transformations

Chapter 1 introduced the isometry transformations that preserve congruence: reflection, rotation, translation, and glide reflection. In our study of similarity in the current chapter, we extend the ideas of transformations to include dilations and the composition of isometries and dilations. A *dilation* is a transformation that maps the plane to itself in such a way that one point O is its own image and any other point P has an image P' on ray OP with $OP' = rOP$, where r is a scale factor. Figure 3.3 depicts a dilation with center O and scale factor 2 that maps $\triangle ABC$ to $\triangle A'B'C'$.

Photo by Robb De Wall. Reprinted by permission of the Crazy Horse Memorial.

U.S. Park Service

Fig. **3.2.**

Mountain sculpture models

(a) Washington head (Mount Rushmore model)   (b) Crazy Horse model (foreground) and mountain (background)

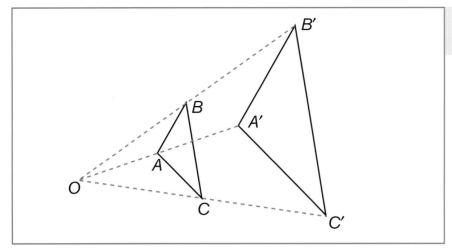

Fig. **3.3.**

Dilation with center *O* and scale factor 2

Principles and Standards calls on high school students to "begin to understand the effects of composites of transformations" (p. 41). An object transformed under the composition of one or more isometries and a dilation results in an image that is similar to the original. The original object's size and shape are preserved through the isometries, and the ratios of lengths or distances are preserved through the dilation. Activities in this section help students move through an informal exploration of dilations, culminating in a challenge to them to determine a composite transformation that relates two similar objects in the plane.

A Copy-Machine Example

The activity Scale Factors allows students to characterize a dilation by working directly with an example that they produce themselves. Students construct a dilation by using a reduction setting on a copy machine, and then they consider the changes that the object undergoes in the copying process. Students begin with a scalene triangle and reduce its sides to 60 percent of their original lengths using a reduction setting of 60 percent on a copy machine. (A reduction setting of 60 percent on a photocopier typically produces an image that is 60 percent of the original size—not an image that reduces the original by 60 percent.) The process is repeated with the new triangle, reducing its sides with a 60 percent setting. At this point, students should have three similar scalene triangles that they can analyze as they respond to questions that help them see significant characteristics of dilations.

The tasks in Scale Factors can also be completed using geometry software. Alternatively, if neither a copy machine nor geometry software is available, students can skip steps 1–3 of the activity and use the triangles in figure 3.4 instead.

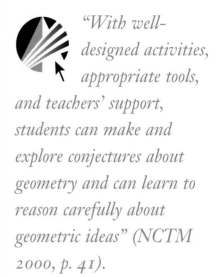

"With well-designed activities, appropriate tools, and teachers' support, students can make and explore conjectures about geometry and can learn to reason carefully about geometric ideas" (NCTM 2000, p. 41).

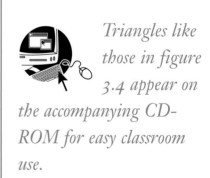

Triangles like those in figure 3.4 appear on the accompanying CD-ROM for easy classroom use.

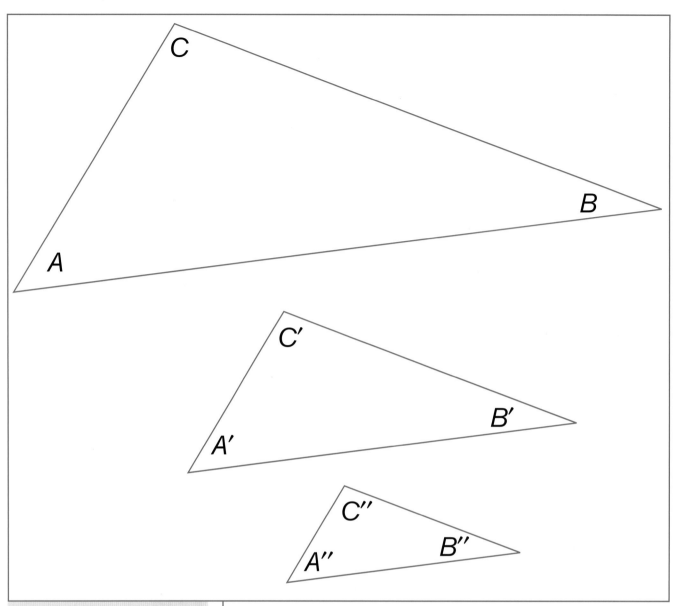

Fig. **3.4.**
Similar triangles

Scale Factors

Goals

- Create similar triangles using copy-machine reductions

- Analyze measurements of lengths among similar triangles created with a known reduction factor.

Materials and Equipment

- A copy of the activity pages for each student

- Access to a copy machine or a computer with drawing software

- A ruler for each student

- Scissors for students to share

- Glue sticks for students to share

- Blank paper

p. 107

Discussion of the Activity

Working through the activity Scale Factors will help students recognize three important characteristics of a dilation. They will see that—

- an object's shape remains unchanged under a dilation;

- lengths associated with the object (sides and distances between points) are reduced, enlarged, or left the same according to some fixed scale factor;

- a point exists from which distances to the object are reduced or enlarged according to the scale factor.

Students draw three lines and find the point of concurrency, which is the point that all the lines have in common. This point is called the *center of dilation*, and from it one figure in the dilation can be mapped to another. Students should have concluded that the ratios of "first reduction" lengths to "original" lengths, such as $A'C'/AC$, and of "second reduction" lengths to "first reduction" lengths, such as $A''P/A'P$, are approximately 0.60, and that ratios of "second reduction" lengths to "original" lengths, such as $A''C''/AC$ and $A''P/AP$, are approximately $0.60 \times 0.60 = 0.36$. Students' values may vary somewhat depending on the accuracy of both their measurements and their placement of the triangles along the three concurrent lines.

The value 0.60 is the scale factor of the dilation. It is no coincidence that the 60 percent copy machine reduction setting matches ratios such as $A'C'/AC$ and $A''P/A'P$. If point P is the center of the dilation, we say that $\triangle ABC$ has been dilated about point P with scale factor 0.60, resulting in the image $A'B'C'$. Using conventional symbolism for dilations, we write that $\triangle A'B'C'$ is the image of $\triangle ABC$ under the dilation $D_{P,0.60}$, or $D_{P,0.60}(\triangle ABC) = \triangle A'B'C'$.

More Dilations

After studying the characteristics of a dilation in Scale Factors, students can complete the activity Basic Dilations to apply what they have learned. They will explore dilations of a quadrilateral and a triangle in this new activity, reinforcing their understanding of the dilation process applied to objects in the plane and giving special attention to the effect of the scale factor on the size of the dilation image.

Basic Dilations

Goals

- Carry out one- and two-step dilations in the Euclidean plane
- Explore the result of a dilation carried out when the center of dilation is within the preimage

Materials and Equipment

- A copy of the activity page for each student
- A ruler for each student
- Geometry software (optional)

p. 110

Discussion of the Activity

In completing the activity and studying the changes that occur with different scale factors (*r*), students discover that for $0 < r < 1$, the image decreases in size; for $r > 1$, the image increases in size; and for $r = 1$, the preimage and the image are congruent. In other words, *a scale factor of 1 is an identity transformation of the plane.* Every point is mapped to itself.

In addition, Discussion and Extension question 3 helps students discover what happens when the center of a dilation is *inside* the preimage. They see that in such a case, they should again consider whether the scale factor *r* is greater than or less than 1. An image under a dilation will be entirely contained within the preimage if $0 < r < 1$. By contrast, the preimage will be entirely contained within the image if $r > 1$.

Coordinating Coordinates

Chapter 1 introduced the use of coordinates in its investigation of transformations. Similarly, a dilation can be investigated with coordinates and matrices. As an introduction to the use of coordinates with dilations, the activity Coordinate Connections allows students to explore how ordered pairs describing the location of an object are transformed under a dilation to create a shape similar to the original.

Like isometries, dilations can also be investigated with coordinates and matrices.

Coordinate Connections

Goals

- Determine the results of dilating objects in the coordinate plane
- Explore the result on a dilation carried out with a negative scale factor

Materials and Equipment

- A copy of the activity pages for each student
- A ruler for each student
- Geometry software (optional)

p. 111

Discussion of the Activity

Discussion and Extension questions for Coordinate Connections encourage students to generalize their results to an arbitrary triangle situated in a coordinate plane. In addition, students are invited to explore the effects of a negative scale factor on a dilation. Students may need prompting to think about extending the dilation through its center when the scale factor r is less than 0. Teachers can use technology to particular advantage here. Geometry software will reveal to students easily and quickly what happens with negative scale factors.

Once teachers have helped students explore dilations in a coordinate plane, it is a natural step to investigating with them how matrices can be used to perform a dilation. Recall that in chapter 1 transformations were written as 3×3 matrices so that they could be multiplied to allow compositions. In Coordinate Connections, students discover that dilating a triangle with the origin as the center of the dilation and a scale factor of 2 amounts to multiplying each coordinate of the vertices of the triangle by 2. They build on that knowledge as they respond to Discussion and Extension questions, discovering that if the origin is the center of a dilation with scale factor r, then any vertex of an original point with coordinates (a, b) has image (ra, rb) under the dilation.

Consider how this notion can be reexpressed with matrices. If the coordinates of the vertex are written in a matrix form, as in chapter 1, they might appear as

$$\begin{bmatrix} a \\ b \\ 0 \end{bmatrix}.$$

Also, their image can be written in matrix form as

$$\begin{bmatrix} ra \\ rb \\ 0 \end{bmatrix}.$$

Asking students how one could generate the image matrix from the original will elicit the inevitable—and in this instance accurate—suggestion: multiply r times the original matrix. It is true that

$$r \cdot \begin{bmatrix} a \\ b \\ 0 \end{bmatrix} = \begin{bmatrix} ra \\ rb \\ 0 \end{bmatrix},$$

but a multiplication of this type does not present a dilation matrix in the 3×3 form of the other transformation matrices.

Now consider what a 3×3 dilation matrix needs to look like. In the example above, each coordinate is multiplied times r, and nothing else changes. In other words, the x-coordinate is multiplied times r, and the y-coordinate is also multiplied times r. Through some experimentation with matrices, students will discover that $D_{(0,\,0),r}$ can be represented by

$$\begin{bmatrix} r & 0 & 0 \\ 0 & r & 0 \\ 0 & 0 & 0 \end{bmatrix}.$$

Equipped with this matrix, students will be prepared to use matrices to compose transformations that include dilations.

Composite Transformations

We know that isometries create congruent figures anywhere in the plane. The activity Multiple Transformations enables students to explore the composition of a dilation with an isometry. Intuitively, students may expect such a composition to produce similar figures that need not be tied in a straightforward or simple way to the origin of a coordinate system. Indeed, they will discover that a dilation followed by an isometry does result in the creation of similar figures anywhere in the plane. Such a transformation is in fact called a *similarity*.

Multiple Transformations

Goals

* Carry out a sequence of transformations that includes one or more dilations

* Create a sequence of transformations that maps an image with a similar preimage

Materials and Equipment

* A copy of the activity pages for each student

* A compass for each student

* A ruler for each student

* Graph paper for each student

* Geometry software (optional)

Discussion of the Activity

Multiple Transformations leads students to explore and extend their understanding of composite transformations. Discussion and Extension questions provide problem-solving opportunities that prompt students to consider the effects of dilations on the perimeter and the area of objects in the plane. Students are invited to make generalizations and build logical arguments about these effects. Question 5, which asks students to speculate why many copy machines have preset image settings of 65 percent, 78 percent, and 129 percent, could serve as an appropriate assessment task for this section of the chapter. Question 6 asks students to design a composite transformation task that includes a dilation. Extending the notion of dilation to three dimensions can provide an additional challenge to students: what happens to lengths, surface area, and volume when a three-dimensional object is dilated through a point?

Similarity Activities

Up to this point, chapter 3 has focused on introducing students to and helping them master procedures required to dilate an object in the plane. We have also explored concepts underlying those procedures. By probing composite transformations that include dilations, we have moved our investigation beyond transformations that preserve congruence to transformations that preserve shape. We have suggested geometry software that can extend the activities as well as additional resources that can expand the study of dilations and similarity. In the next section, we will examine applications of similarity.

Principles and Standards proposes that students in grades 9–12 should be able to "use geometric models to gain insights into, and answer questions about, other areas of mathematics" (p. 308). We now present two problem-solving situations that integrate data col-

p. 113

For additional work with similarity, dilations, and matrices, see pages 55–56 in Geometry from Multiple Perspectives *(NCTM 1991). The authors begin with coordinate and matrix representations, which they see as "an especially convenient vehicle for introducing similarity and the related transformations"* (p. 55).

lection and analysis with concepts of similarity and procedures for applying them. Activities in this section are grounded in indirect measurement applications.

Shadows and Lengths

The activity Shadowy Measurements offers a slightly different twist on the classic problem of making an indirect measurement using similar triangles and shadow lengths. Students explore the relationship between the heights of a collection of objects and the lengths of the shadows created by those objects at a particular time on a sunny day. Students collect data, examine them, and apply the relationships they find to determine the unknown height of a tree, building, or similar object.

Shadowy Measurements

Goals

- Collect, organize, and analyze data to determine a direct proportion between objects' heights and their shadows' lengths
- Use a direct proportion to estimate an object's height on the basis of the known length of its shadow
- Interpret the data collected in relation to similar triangles

Materials and Equipment

p. 115

- A copy of the activity pages for each student
- Metersticks, yardsticks, or measuring tapes

Note: The activity has been designed for an outdoor setting, but teachers and students can work indoors using an alternative setup that involves an architect's lamp with a bright bulb, a group of objects of known and unknown heights, and a yardstick.

Discussion of the Activity

At a given time of day, rays of sunlight strike objects in a particular area of the earth's surface at virtually the same angle. Shadows created by the sun's rays create similar triangles, as depicted in figure 3.5. The similar triangles ensure that each ratio,

$$\frac{\text{shadow length}}{\text{object height}},$$

is the same. Table 3.1 illustrates the sorts of measurements that students might find when they complete the activity.

Table 3.1

Sample Data Values for Shadowy Measurements

Object	Length of Shadow	Height of Object
Fence post	6.4 feet	8 feet
Basketball backboard (top to ground)	8.8 feet	11 feet
Bicycle tire	1.6 feet	2 feet
Car radio antenna (tip to ground)	4.4 feet	5.5 feet
Yardstick	2.4 feet	3 feet

The relationship between "length of shadow" and "height of object" in the table, as well as in the data that students collect, is a direct proportion. As a result, the points on the students' scatterplots of their data ought to be located along a straight line containing the origin. A scatterplot of the data from table 3.1 and the line through the points are shown in figure 3.6.

Students can use the scatterplots that they create in Shadowy Measurements to estimate any object's height by finding the point along the line that corresponds to the length of its shadow. They may need help in realizing that the relationship holds only when the measurements have been made at approximately the same time of day within the same general locality.

Some students may be bothered by the fact that the graph plots "object height" against "shadow length." However, students will find that it is both instructive and meaningful to construct the graph in this manner. The shadow lengths and the heights of the given objects in table 3.1 can be measured directly, as can the length of the shadow of a tall object. In this situation, at the particular time when the data were collected, the shadows can be considered as the independent variables for the graph. When the graph is plotted in this manner, the slope of the line determined by the data points is the scale factor of the similar figures—certainly not an insignificant result.

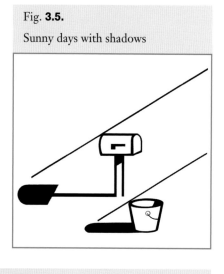

Fig. **3.5.**

Sunny days with shadows

Using Shadow Lengths to Predict an Object's Height

Fig. **3.6.**

Shadow graph

Peering through a Tube

Most people at one time or another have used a cardboard or plastic tube as a telescope. Although the tube does not actually enlarge what we see, it helps us focus on a narrow field of vision. The activity Field of Vision reinforces and extends ideas about similarity that students explored in Shadowy Measurements, allowing them to examine relationships that may exist among a tube's length, its diameter, the viewer's distance from an object, and his or her field of vision through the tube.

Discussion and Extension question 3 asks students to undertake some research on the ancient Greek astronomer Eratosthenes, who used shadows and proportions to estimate the earth's circumference. Web sites that can provide useful information are www-history.mcs.st-and.ac.uk /~history/Mathematicians /Eratosthenes.html and www.eso.org/outreach /spec-prog/aol/market /collaboration/erathostenes.

Field of Vision

Goal

- Collect, organize, and analyze data to determine a relationship between the characteristics of a viewing tube and its field of vision

Materials and Equipment

- A copy of the activity pages for each student
- Metersticks or measuring tapes
- Viewing tubes of various sizes (for example, paper towel rolls and PVC pipes)
- Masking tape
- A graphing calculator for each student (or graphing software)

p. 117

Discussion of the Activity

In experimenting with viewing tubes of various diameters and lengths, students will be applying what they have learned about similar triangles to the relationships among the diameter of a tube, its length, the viewing distance, and the field of vision. Providing students with a number of congruent tubes will enable them to compare their results directly.

Tubes from paper for fax machines and digital duplicators may be readily available from the school's print shop or a benevolent small-business owner in the community. Other workable tubes include rolls from paper towels, toilet paper, plastic wrap, aluminum foil, and gift wrap. Plastic pipe (PVC) of various diameters and lengths can also be used. If no tubes are available, students can make viewing tubes from cardboard or construction paper.

Teachers can use figure 3.7 to illustrate the variables of the situation and highlight the similar triangles that are important for analyzing it. The isosceles triangles *ECB* and *EXY* are similar because they share angle *E* and *CB* and *XY* are parallel to each other. As a result of their similarity, we know that the lengths *d* (diameter of the tube), *t* (length of the tube), *f* (diameter of the viewer's field of vision), and *w* (distance from the viewer's eye to the wall) are in the relationship: $d/t = f/w$. If we solve this equation for *f*, the field of vision, we get

$$f = \frac{wd}{t}.$$

This equation sets the three lengths that we can determine directly—the viewing tube's length, its inside diameter, and the distance from the viewer's eye to the wall—in a relationship that is equal to the measure we seek—the viewer's field of vision.

Dynamic geometry software such as Geometer's Sketchpad or Cabri Geometry II can be used to model the setting of Field of Vision. This software can enable users to explore the problem by altering any of the three known lengths (*d*, *t*, or *w* in the figure). With this software, users can also create a sketch in which they can vary the position of the viewer's eyes, moving them up or down, and thus they can do away with

Wilson and Shealy (1995) present similar data-collection activities that focus on functional relationships and mathematical connections. See the CD-ROM that accompanies this book.

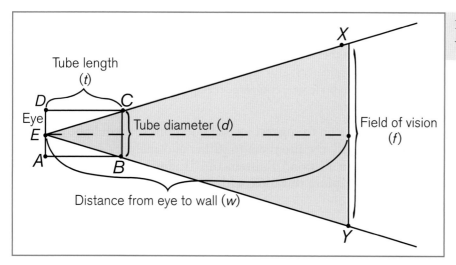

Fig. **3.7.**

Viewing tube geometry

the activity's assumption that the viewing tube is perpendicular to the wall.

An applet that allows users to drag elements of figure 3.7 is available on the accompanying CD-ROM.

Taking Similarity Another Step

The two preceding activities, Shadowy Measurements and Field of Vision, briefly illustrate indirect measurement determined by ratios and similarity. Teachers who wish to do so could introduce a related, potentially stimulating activity that highlights historical or factual information about famous people while teaching students about the geometry of similar figures. Teachers would direct students' attention to a historical person's three-dimensional likeness that has been cast in stone, metal, or another material. They would give their students the measure of one feature in the likeness. For example, they might tell students the measure of Lincoln's mouth as carved on Mount Rushmore, the length of Crazy Horse's nose as it appears on the mountain tribute to him, or the length of the Statue of Liberty's arm in New York City. Teachers would then ask students to devise a strategy for collecting and analyzing data to determine another measure related to the first one, such as the carved length of Crazy Horse's outstretched arm or Lincoln's hypothetical height if his likeness on Mount Rushmore showed him in full, from head to toe.

This type of data collection focuses on linear measures. Once students become adept at the process, questions could be posed about one-, two-, or three-dimensional characteristics of the human body. Working from J. B. S. Haldane's article "On Being the Right Size" (1956), MacPherson (1985) describes a sequence of such similarity activities:

> Once the three basic rules for linear measures, areas, and volume are established, the fun starts. Begin with animals' legs. The weight of an animal depends largely on its volume. But the strength of its legs depends primarily on their cross-sectional area. So if we double the size of an animal (and by that we mean all linear measures), its weight will be increased by a factor of eight ($2 \times 2 \times 2$), but the strength of its legs will increase by a factor of only four (2×2). Its legs will not be able to

Students could be asked if human beings could possibly be as large as the Statue of Liberty—or, for that matter, the Jolly Green Giant. Could such beings exist in the real world?

support its weight so readily. As we increase the size yet further, the animal soon comes to the point where its legs can no longer support its body. That, together with some other reasons to follow, is why there are no house-sized spiders. It also explains how quite large animals do cope. In proportion to body size, compare an elephant's legs with a spider's. (P. 77)

MacPherson goes on to suggest several questions that could be posed to students and would require them to gather and analyze data. In "Godzilla: Fact or Fiction," Billstein and Trudnowski (1989) offer another way to explore this topic with students.

Conclusion

This chapter has examined similarity in various representations and aspects. Teachers and students have looked at similar figures through the lens of transformations, with and without coordinates and matrices. They have participated in an examination of a variety of applications of similarity in the real world. Throughout the process, they have undoubtedly heard our persistent refrain: Geometrical topics do not exist in isolation and should not be studied that way. Geometry can be used as a tool to study other topics in mathematics, and vice versa. With this chapter as background, readers should be ready to experience the extension into the geometric infinite that chapter 4 offers.

Help students consider other two- and three-dimensional possibilities with "Godzilla: Fact or Fiction" by Rick Billstein and James Trudnowski on the CD-ROM that accompanies this book.

NAVIGATING *through* GEOMETRY

Chapter 4
Visualizing Limits in Our World

Students' early experiences with concrete patterns make their later investigations of cases involving infinity more credible and meaningful to them.

Many mathematical ideas are introduced to students through visual representations. *Principles and Standards for School Mathematics* (NCTM 2000) recognizes the power of visual models to cement and express geometrical understanding, calling for all students in grades 9–12 to be able to "use visualization, spatial reasoning, and geometric modeling to solve problems" (p. 308) and to "understand relations and functions and select, convert flexibly among, and use various representations for them" (p. 296). In addition, *Principles and Standards* expects all students in grades 9–12 to be able to "use geometric models to gain insights into, and answer questions in, other areas of mathematics" (p. 308).

With these Standards and expectations in mind, consider students' explorations of patterns from the earliest grades. Students typically start with concrete or visual patterns, which are customarily finite. However, these experiences are not unconnected to their later explorations of more complex or abstract patterns. Without a preliminary understanding of finite patterns, students would have little understanding of repeating decimals, irrational numbers, or arguments about estimating π or finding the area of a circle. To aid teachers in keeping the Standards in view while they motivate students to think about the infinite, we now draw on the mathematics of the first three chapters to consider another major application of geometry—the use of geometric models to investigate concepts that include notions of the infinite and provide an introduction to limits. Such a use of geometry pervades high school mathematics and is assumed, usually without discussion, in textbooks' presentations of many concepts.

In this chapter, we look first at geometric approaches to sequences and series as natural extensions of the patterns that students begin to examine in the early grades, then we consider the application of similarity to fractals and measurements involving fractal geometry, and afterward we examine a limiting function from the geometry of a kaleidoscope.

Sequences and Series

Sequences and series are two sophisticated and important mathematical concepts. Students usually encounter sequences and series first in finite examples, but they can also think intuitively about examples that involve infinity by means of visual representations. Any discussion of sequences and series inevitably leads to the idea that an infinite sum of numbers may or may not "add up," or *converge*, to a finite number. It is not difficult for students to see intuitively that the sum $1 + 1 + 1 + 1 + 1 + 1 + 1 + 1 + \ldots$ of an infinite number of 1s is an infinite number or that the expression $1 - 1 + 1 - 1 + 1 - \ldots$ has no sum. However, students need to be able to reason at some level of abstraction to visualize the infinite processes involved even in these simple examples. Most students find it much more difficult to see intuitively that the sum of the nonnegative powers of 1/2 is a finite number.

In this section, we try to stimulate students' mathematical intuitions by using geometrical models in such a way that meaningful change may occur in their mathematical understandings. The initial activities, What's My Sum? and Sum Me Up, illustrate this approach.

Geometric Sequences and Series

An infinite geometric *sequence* is an ordered set of terms $a_1, a_2, a_3, \ldots, a_n, \ldots$, indexed (or counted) by the natural numbers, where $a_n = ra_{n-1}$, for $n \geq 2$ and ratio r. For example, 1, 2, 4, 8, 16, 32, … , the list of all nonnegative powers of 2, is an infinite geometric sequence with ratio 2. An infinite geometric *series* is a sequence $s_1, s_2, s_3, \ldots$, where s_n is the sum of the first n terms of an infinite geometric sequence. For example, 1, 3, 7, 15, … is the infinite geometric series formed from the infinite geometric sequence 1, 2, 4, 8, … , which contains the nonnegative powers of 2.

In an article provocatively named "Mathematics without Words," Mabry (2001) presents the sum of an infinite geometric sequence in a picture similar to the one in figure 4.1. The sequence begins with 1, and each succeeding term is found from the previous one by multiplying by 1/3. Mabry's picture demonstrates the same thing that students discover in a different way in the activity What's My Sum?

Mathematics books use the word series *in different ways. It is used in analysis as we have defined it here, as a sequence of partial sums. However, it is also used in many places as the limit of the sequence of partial sums.*

Fig. **4.1.**

A geometric interpretation of the infinite sum $1/3 + 1/3^2 + 1/3^3 + 1/3^4 + \ldots = 1/2$

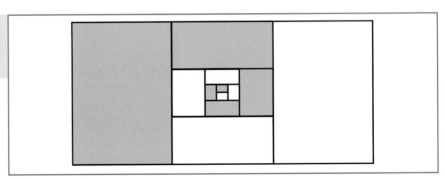

What's My Sum?

Goals

- Construct a visual representation of an infinite geometric series
- Understand the concept of an infinite series
- Understand the ideas of convergence and divergence of a geometric series
- Conjecture about the convergence or divergence of a series

Materials and Equipment

- A copy of the activity pages for each student
- A sheet of paper (8 1/2 × 11 in.) for each pair of students
- Scissors for each pair of students

Discussion of the Activity

p. 119

In this activity, students "construct" the geometric sequence 1/3, 1/9, 1/27.... To do this, they cut a sheet of paper into thirds, keeping one third in their hands and placing the other two thirds on a desk or table to form the beginnings of two separate piles. They then repeat the process with the paper left in their hands, cutting it into thirds and again keeping one third and stacking the other two on the piles on the table. Because the piece of paper in their hands at the beginning of the second step was one-third of its original piece, they will be holding one-ninth of the original at the end of this step. Continuing this process until the paper in their hands is virtually gone, the students will see that each pile on the table must contain half of the original sheet of paper. Thus, they have a visual representation—actually, two duplicate representations—of the infinite sum of $1/3 + (1/3)^2 + (1/3)^3 + (1/3)^4 + \ldots + (1/3)^n$ *as* 1/2.

Looking at the Algebra

The sum of a geometric sequence can also be investigated algebraically. This is accomplished in the activity Sum Me Up.

Sum Me Up

Goals

- To investigate the algebra of an infinite geometric series
- To find a formula for the sum of a convergent geometric series

Materials and Equipment

- A copy of the activity page for each student
- A spreadsheet (optional)
- A computer algebra system (optional)

p. 121

Discussion of the Activity

Sum Me Up asks students to use algebraic techniques for polynomial division to find the quotient of $(1 - x^{n+1}) \div (1 - x)$. Their division will yield the sum $1 + x + x^2 + \ldots + x^{n-2} + x^{n-1} + x^n$. After students have substituted $1/2$ for x, as the activity directs, they can see that what they have is an expression for the sum of the first n terms of the infinite geometric series of powers of $1/2$ (starting with the zeroth term, 1). Recognizing that this is true, they will be ready to consider what happens to both the value of $(1/2)^n$ and the sum of the series as n becomes infinitely large.

Other Geometric Series

As an extension of the activities What's My Sum and Sum Me Up, we could use other geometric shapes to visualize different geometric series. For example, to visualize the sum $(1/4) + (1/4)^2 + (1/4)^3 + (1/4)^4 + \ldots + (1/4)^n$ as n goes to infinity, we could have students consider shaded triangles such as those shown in figure 4.2. If the original triangle has area A, then the shaded triangles have a total area that is equal to one-third of A. If the area of A is 1, then we have

$$(1/4) + (1/4)^2 + (1/4)^3 + (1/4)^4 + \ldots = (1/3)(1) = 1/3.$$

To reproduce this effect another way, we might modify the activity in What's My Sum? by having students begin with an equilateral triangle that they have cut from paper. They could then cut the triangle into four congruent equilateral triangles, each with an area equal to one-fourth of that of the original (see fig. 4.2). Now they could follow the same process as before, except that this time they would be making three piles of triangles, instead of two piles of rectangles, and keeping the fourth triangle in their hands. Again, they would repeat the process with the paper that they were holding until they had virtually no paper left. The process would enable them to visualize what fraction of the original triangle was contained in each pile.

This geometric visualization would exemplify the generalized result that the series formed from a geometric sequence with first term a and ratio r, where $|r| < 1$, is the infinite sum

Fig. **4.2.**

Equilateral triangle with shading

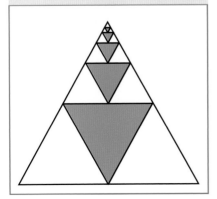

Navigating through Geometry in Grades 9–12

$$a + ar + ar^2 + ar^3 + \ldots + ar^n + \ldots = a\left(1 + r + r^2 + r^3 + \ldots + r^n + \ldots\right)$$

$$= a\left(\frac{1}{1-r}\right) = \frac{a}{1-r}.$$

The geometric series explored in What's My Sum? and Sum Me Up is formed from the infinite geometric sequence that has 1/4 as both its first term and its ratio, and the infinite sum is

$$\frac{1}{4} + \left(\frac{1}{4}\right)^2 + \left(\frac{1}{4}\right)^3 + \ldots + \left(\frac{1}{4}\right)^n + \ldots = \left(\frac{1}{4}\right)\left[1 + \left(\frac{1}{4}\right) + \left(\frac{1}{4}\right)^2 + \ldots + \left(\frac{1}{4}\right)^{n-1} + \ldots\right]$$

$$= \left(\frac{1}{4}\right)\left(\frac{1}{1 - \left(\frac{1}{4}\right)}\right) = \left(\frac{1}{4}\right)\left(\frac{4}{3}\right) = \frac{1}{3}.$$

The use of visual geometry to examine applications involving infinity continues in the next section on fractals.

Geometry with Fractals: The Koch Snowflake Curve

In 1983, Benoit B. Mandelbrot described his work on a new topic in geometry:

> I conceived and developed a new geometry of nature and implemented its use in a number of diverse fields. It describes many of the irregular and fragmented patterns around us, and leads to full-fledged theories, by identifying a family of shapes I call *fractals*. The most useful fractals involve chance and both their regularities and their irregularities are statistical. Also, the shapes described here tend to be *scaling*, implying that the degree of their irregularity and/or fragmentation is identical at all scales. (Mandelbrot 1983, p. 1)

Mandelbrot drew attention to the importance of both similarity and infinite processes in the study of fractals. A property of some fractals is *self-similarity*. Informally, self-similarity in fractals ensures that one can view a fractal image from a variety of perspectives—up close, far away, or under a microscope—and see the same shape at every level. In this section, we explore the concept of self-similarity and some infinite series that arise naturally in the study of fractals.

We can readily see self-similarity in the Koch snowflake curve, which was introduced by Swedish mathematician Helge von Koch in 1904. The curve, shown in figure 4.3a, can be drawn in a variety of ways. One way is to break it into smaller pieces and draw it in parts. Figure 4.3b shows that the snowflake could be made of three congruent pieces of a curve, appropriately placed. Each of the three pieces could be considered a fractal.

Figure 4.4 shows how we could start the Koch snowflake curve with a line segment. To begin the iterative process that ultimately creates the

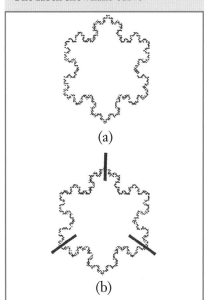

Fig. **4.3.**

The Koch snowflake curve

(a)

(b)

Fig. **4.4.**

Initiating the fractal

The Koch Snowflake Curve applet on the accompanying CD-ROM lets users experiment with the curve's iterative process.

fractal image, we remove the middle third from the segment and replace it with two segments each of whose length is one-third the length of the original, as shown. (A different way to draw the curve is to start with an equilateral triangle and create the "bumps" as described, in its sides. This is the method used in the activity called The Koch Snowflake Curve: How Big Am I?)

Each pair of new segments forms part of an equilateral triangle in the middle of the side of the original segment. An object that might look something like a part of a star is created. The new shape transforms the original segment into four segments, each of which is one-third as long as the original.

We continue to apply this modification, removing the middle third from each segment and replacing it with two segments, each one-third of the length of the segment that we just divided. From four segments in the second step, we next generate sixteen segments, as shown. This iterative process is continued infinitely, generating a fractal image that exhibits self-similarity. That is, the smallest sections of the curve are segments that are clearly similar to the original segment. Also, any small, starlike section is similar to the section with the original "bump." Once this curve has been created, then copies of it can be assembled to form the snowflake curve of figure 4.3.

The eye-catching Koch snowflake curve displays a perimeter with a vast array of fractured, twisting segments, all resulting from the repeated application of a seemingly innocent shape modification. In the following activity, The Koch Snowflake Curve: How Big Am I? students investigate the perimeter of the snowflake curve using a spreadsheet and discover that its perimeter provides an example of an infinite geometric series. (The activity includes special contributions from Collin Joyce of Freemont, California.)

The Koch Snowflake Curve: How Big Am I?

Goals

- Investigate the perimeter and area of the Koch snowflake curve
- Conjecture whether the perimeter of the curve is finite or infinite
- Conjecture whether the area enclosed by the curve is finite or infinite
- Verify conjectures algebraically
- Gain an intuitive understanding of the idea of a limit

Materials

- A copy of the activity pages for each student
- Spreadsheet software

p. 122

Discussion of the Activity

The activity The Koch Snowflake Curve: How Big Am I? investigates how the area and the perimeter of the snowflake grow as the number of iterations of the fractal-building step increases. In particular, students explore whether the area and perimeter are finite or infinite quantities. A spreadsheet aids their investigation, allowing them to explore mathematical formulas that generate area and perimeter while they look for a general pattern. After several steps of the investigation, they can generalize to see that the snowflake has a finite area but an infinite perimeter.

The activity includes extensions to three dimensions, as described by Collin Joyce when he was a high school student in Fremont, California. In an unpublished paper entitled "A 3-D Analog to the Snowflake Problem" (2000), Joyce suggests that the three-dimensional analog of the snowflake curve can be investigated by high school students. He writes, "If a regular tetrahedron has a regular tetrahedron 'grown' on each face and the process is repeatedly replicated, is there a limiting surface area and a limiting volume? If so, what are they?" Figure 4.5 shows this progression of solids. In figure 4.5b, the regular tetrahedron shown in figure 4.5a has grown a regular tetrahedron (with a side half as long) on each of its faces. Figure 4.5c shows this new solid in the process of growing tetrahedra on its faces in the same way. These solids, whose limiting structure is a cube, are considered in the Discussion and Extensions questions that accompany the activity.

A Limit in a Geometric Application

Most functions that we study and that arise naturally are continuous. Intuitively, we are persuaded that a continuous function has a graph that may be drawn without lifting the pencil from the paper. However, discontinuous functions also occur naturally. Students investigate one

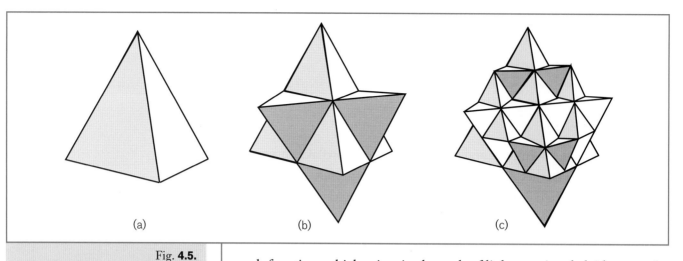

Fig. 4.5.

Three-dimensional snowflake figures: (a) a regular tetrahedron; (b) a regular tetrahedron that has "grown" a regular tetrahedron on each face; and (c) the solid shown in (b) in the process of "growing" tetrahedra in the same way.

such function, which arises in the path of light rays in a kaleidoscope, in the activity *Dis*continuous, That's What You Are!

In chapter 1, we saw that a rotation can be written as the composition of two reflections over nonparallel lines. This central principle of a kaleidoscope is investigated in both of the following activities, Smoke and Mirrors and *Dis*continuous, That's What You Are! The first one, Smoke and Mirrors, serves as a general introduction to kaleidoscopes.

A conventional triangular kaleidoscope is composed of three rectangular mirrors joined to one another along the longer edges to form the sides of a triangular prism. Light inside the prism is reflected by the mirrors, forming interesting patterns from the various items—often, pieces of colored glass or colored plastic shapes—that are placed on the end of the prism. In Smoke and Mirrors, students explore the reflecting properties (and symmetry) of two rectangular mirrors joined along one edge. (This activity is based on a project found in *Discovering Geometry: An Inductive Approach* [Serra 1997, p. 394].)

Smoke and Mirrors

Goals

- Understand principles of light reflection
- Understand principles of a kaleidoscope
- Investigate what happens when the angle between mirrors in a kaleidoscope is changed

Materials and Equipment

- A copy of the activity page for each student
- Two rectangular mirrors for each student or group of students
- Tape
- Patterned paper or gift wrap
- Plain paper
- A protractor for each student or group of students

p. 125

Discussion of the Activity

By varying the angle between two rectangular mirrors taped together, students see different patterns emerging in reflections of a printed piece of gift wrap. Then they place the hinged mirrors on a piece of plain paper with a simple object on it like a penny or a line segment that they have drawn, and they can observe the number of images created by different angles.

The Geometry of a Discontinuous Function

In the next activity, *Dis*continuous, That's What You Are! students look at a simple mathematical representation of the path of a light ray in a conventional three-sided kaleidoscope. They model the light as a ray inside an equilateral triangle whose sides have the reflective properties of mirrors.

Suppose the light enters the kaleidoscope at a point on one side and continues along a path that is parallel to another side of the kaleidoscope, as in figure 4.6. When it reaches one of the other sides, the light ray is reflected toward the third side, with the angles of incidence and reflection being equal. After the first "bounce," the path of the light becomes parallel to the side where the ray entered. The ray of light continues traveling and bouncing in this way inside the kaleidoscope. It is reasonable to ask how far the light travels.

Fig. **4.6.**

Kaleidoscope problem

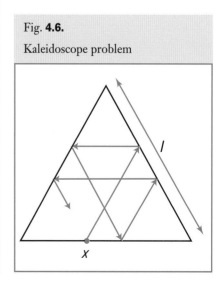

*Dis*continuous, That's What You Are!

Goals

p. 126

- Investigate the path of light inside a kaleidoscope
- Graph the path of the light as a function of its point of entry
- Understand that discontinuous functions occur naturally

Materials

- A copy of the activity page for each student
- Plain paper and graph paper or isometric dot paper for each student
- Geometry software (optional)

Discussion of the Activity

In the activity, students draw line segments in an equilateral triangle to represent light rays in a kaleidoscope. Then, using familiar ideas about parallelograms and equilateral triangles, together with the principle that the angle of incidence equals the angle of reflection, they measure the segments. In this way, they determine lengths of different paths of light. The process enables them to discover that the path of light rays inside a three-sided kaleidoscope is an example of a discontinuous function that originates in an everyday setting.

Students determine that the light travels a distance that is equal to three times the length of one side of the triangle *unless* the light enters exactly at the midpoint of a side. They find that in this case the light travels only one and a half times the length of one side. A graph of the distance traveled by the light as a function of the position at which it entered the kaleidoscope represents a naturally occurring discontinuous function (Cofman 1990).

To consider the discontinuity more formally, let $f(x)$ be the function that describes the distance traveled by the light entering the three-sided kaleidoscope in figure 4.6 at a given starting point x. Let the origin be at the lower left vertex of the triangle and l be the length of a side. The domain of f is the set $0 \leq x \leq l$. Using the familiar geometry of triangles and parallelograms, we can show with little difficulty that $f(x) = 3l$ except when $x = l/2$, in which case $f(l/2) = 3l/2$. This follows from our supposition that the light entered parallel to a side and that each segment representing the path of the light was parallel to one of the sides of the triangle. Using the property of parallelograms that opposite sides are congruent, the result follows. In other words, the graph of f is discontinuous at $x = l/2$. The value of f as x approaches but is not equal to $l/2$ is $3l$, even though the value of f when $x = l/2$ is $3l/2$. Mathematically, we write

$$\lim_{x \to \frac{l}{2}} f(x) = 3l,$$

and we read this as "the limit of $f(x)$ as x approaches $l/2$ equals $3l$," whereas $f(l/2) = 3l/2$.

A Software Limitation

This function raises an interesting and important point about geometry software. Because geometry utilities use finitely many points to approximate the real plane, a utility may "skip over" and thereby miss something important and possibly unique that is happening at a particular point. Thus, a discontinuity that occurs in a function at only one point—in the case of our function, the midpoint—may not reveal itself in collected data. In simulating the function that we are considering here, some types of technology may give the value of $f(x)$ when $x = l/2$ as $3l$. Teachers should be aware of this flaw in the technology's method and be prepared to discuss it with students.

The kaleidoscope illustrates how a "real" geometry problem—one that arises innocently in the everyday world—can be investigated with traditional geometric ideas. It also presents an example of how a discontinuous algebraic function can be examined in a geometric setting. In addition, it gives teachers an opportunity to talk about the shortcomings of technology and the dangers of reliance on it without a full understanding of the mathematics involved in particular situations. Finally, it allows students to determine the limit of a function through geometric visualization.

Our next example, which happens also to be our last, draws on data analysis for yet another situation where geometric visualization helps to explain a mathematical concept—this time, a concept that is commonly used in statistics.

Visual Display of Squares in "Least Squares Line of Best Fit"

The method of least squares is a familiar concept to many students, and the algorithm is often applied through the use of technology that finds a line that "fits" given data. However, the explanation of how this algorithm works and why it is a viable way to find a line that represents the "best fit" for a set of data is not so well understood. The advent of dynamic statistics and geometry software makes an investigation by means of a visual representation of the "sum of the squares" both worthwhile and understandable. Students can see the true meaning of the line-of-best-fit method by viewing the line along with the set of data points that it seems to fit and drawing squares that show the amount of error in the fit.

Dynamic statistics or geometry software allows us to graph a set of data easily and to "eyeball" a line that we think fits it. At the same time, we can display the squares that show the error of the fit, as illustrated in figure 4.7. (Note that the data shown in the figure are samples from the activity How Small Are the Squares?) The area of each square represents the square of the distance between the data point and the

Dynamic statistics and geometry software lets students see what the least squares line of best fit really means.

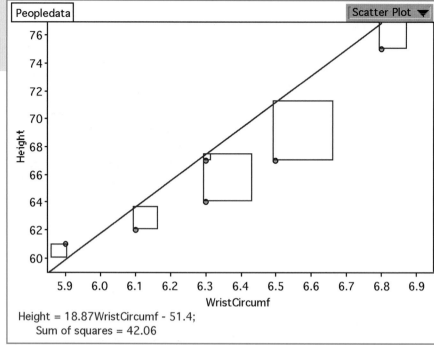

Height = 18.87WristCircumf - 51.4;
   Sum of squares = 42.06

estimated line. Using the dynamic features of the software, we can move the line until the area of the squares appears to be minimized. Then we can graph the actual line of best fit and see how close we came to it. We can also see the error in the "best fit"—that is, we can look at the minimum sum of the areas of all the squares (Finzer, Erickson, and Binker 2000; Vonder Embse 1997).

In the activity How Small Are the Squares? students collect a small set of data, such as the heights and wrist circumferences of their classmates, plot these variables, and find a least squares line of best fit.

Teachers who do not have access to a statistics software package can use the regression line applet at the Illuminations Web site: illuminations.nctm .org/imath/912 /LinearRelationships/index .html#first. The applet is also available on the accompanying CD-ROM, as is Charles Vonder Embse's article "Visualizing Least-Square Lines of Best Fit" (1997).

How Small Are the Squares?

Goals

- Learn to use a dynamic statistics package or a dynamic graphing utility to investigate relationships among data
- Approximate the "least squares line of best fit" by experimenting with a movable line
- Display the squares that the algorithm seeks to minimize and experiment to make them as small as possible
- Compare the approximation to the actual "line of best fit" provided by the utility
- Discuss and explain the meaning of the relevant terminology

Materials and Equipment

- A copy of the activity page for each student
- Dynamic statistics package or other dynamic graphing utility

Discussion of the Activity

Students are urged to come up with two variables that may be correlated in some way, such as the heights and wrist circumferences of their classmates. Then they use dynamic geometry software to plot one variable against the other—for example, wrist circumference against height. Using the "movable line" feature of dynamic statistics software, they impose a straight line on the graph of their data and move it around until they have it positioned where it appears to fit the data better than it did anywhere else. Using the utility's "show squares" feature, they then display the squares that visually represent the error between the chosen line and the actual data. By moving the line, they observe the changing value of the sum of these squares and can then attempt to position the line so as to minimize this sum—that is, to find the "least squares line of best fit," as seen in figure 4.8.

After experimenting with their line, they can have the software draw the actual line of best fit. These two graphs are superimposed in the graph window, with the values of the sum of squares for the hand-picked and the software-generated lines shown simultaneously, along with the actual squares. The screen in figure 4.8 shows the result and the regression equation for each line.

p. 127

The accompanying CD includes a set of general directions for Sketchpad users for setting up the geometry picture for the least squares.

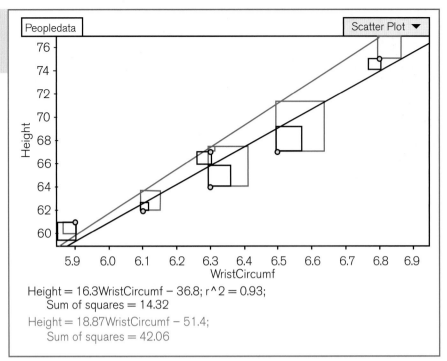

Height = 16.3WristCircumf − 36.8; r^2 = 0.93;
   Sum of squares = 14.32

Height = 18.87WristCircumf − 51.4;
   Sum of squares = 42.06

The activity brings new meaning and relevance to the concept of "least squares line of best fit." Students gain a clear picture that assists their understanding of how to fit the "best" line—that is, the line yielding the smallest square-area sum—to a data set.

Conclusion

This chapter has presented examples of geometry that extend to notions of the infinite in mathematics. Students and teachers have looked at pictures that let them "see" the limits of particular infinite series, and they have made these types of pictures themselves. In considering ideas of fractals in both two and three dimensions, they have encountered geometric topics that are not customarily addressed in high school courses. In the process, they have experienced the unique power of geometry to make abstract ideas highly visual and concrete.

NAVIGATIONS
SERIES

GRADES 9–12

NAVIGATING *through* GEOMETRY

Looking Back and Looking Ahead

The geometry of Euclid is neither the geometry of *Principles and Standards for School Mathematics* (NCTM 2000) nor the geometry of this book. As we have seen, geometric ideas range from a transformation approach to Euclidean geometry to the use of coordinates and technology in investigating topics not only in geometry but also in other fields of mathematics. These approaches could not have been implemented in the time of Euclid, since neither the mathematical knowledge nor the technology was advanced enough to support them.

Principles and Standards provides an opportunity for healthy change in traditional geometry. Robust mathematics encourages connections among diverse topics and allows for an algebraic approach, including the use of matrices, in studying geometry. Expanding from the 2×2 matrices often used in high school mathematics to 3×3 matrices yields a more mathematically structured treatment of transformation geometry. Similarly, expanding from points considered as locations in a plane, with Cartesian coordinates, to points considered as locations in the real world, with coordinates of latitude and longitude, permits classroom investigation of the advanced geometry behind the technology of the global positioning system (GPS). Finally, expanding from finite geometrical patterns and processes to infinite ones in the study of fractals gives students a compelling visual introduction to the idea of limits.

We hope that this book will encourage us all—readers and authors alike—to sustain the openness of mind that we have attempted to cultivate here and that this spirit will continue to shape our thinking as we continue to consider what geometry may become in the future.

NAVIGATIONS SERIES

GRADES 9–12

NAVIGATING *through* GEOMETRY

Appendix

Blackline Masters and Solutions

Fold Me! Flip Me!

Name _____

1. On a sheet of paper, draw any object, such as the flower pictured here, placing it away from the center of the page. Mark any point on your picture with the letter *A*.

2. Fold the paper in half with the picture facing out, and trace it on the blank side of the folded paper. Your new image will be opposite your original picture when you unfold the paper. Mark the image of *A* as *A′*. Draw a line along the fold line, and draw $\overline{AA'}$.

3. Describe the relationship between the fold line and $\overline{AA'}$.

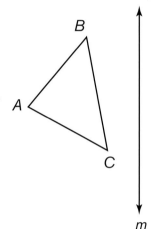

4. Place a Mira or similar tool along the fold line and check the reflection image of your picture.

5. Use a different piece of paper. Draw △*ABC* and a reflecting line *m*. Reflect the triangle in line *m* to find the image △*A′B′C′*.

   a. How is line *m* related to each segment of $\overline{AA'}$, $\overline{BB'}$, $\overline{CC'}$?

   b. Given any point *P* in the plane of the paper, describe how you can find its image *P′* under a reflection in line *m*.

   c. Given any point *P′*, describe how you could find its preimage, point *P*, in a reflection in line *m*.

   d. Given any point *P′*, how could you find point *P* if you had to use a compass and straightedge only?

6. Given the fragment of an ancient pottery plate shown at the right, how could you find the center of the whole plate?

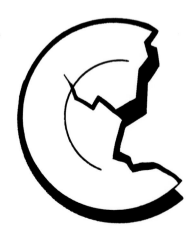

Fold Me! Flip Me! (continued)

Name _____

Discussion and Extension

1. Informally, the *orientation* is the direction you would walk around a figure following its outline. For the reflections in the activity, describe the orientation of a figure's image in relation to the original figure (or preimage).

2. Suppose a reflection is used to find the image of a figure, and then you reflect that image in another line. Describe the orientation of the final image compared with the orientation of the original.

3. Generalize your findings for Discussion and Extension questions 1 and 2.

4. Describe a circle using reflecting lines and the images of the circle.

Mirror, Mirror, on the Wall

Name _____

1. The picture shows Polygon, a character from the Figure This! Web site (www.figurethis.org). Use a straightedge (or the edge of a Mira or similar tool) to draw a line segment from Polygon's left eye to the place on the mirror where the top of the image of her hat appears. Draw another segment from the top of the image of the hat to the top of Polygon's "real" hat.

2. Draw a vertical line from the top of the character's hat to the bottom of her feet.

3. Construct a perpendicular to the vertical line drawn in step 2 that goes through the top of the hat in the image. The perpendicular will form angles with the segments drawn in step 1. How do the measures of these angles compare to each other? The angles are known as the *angle of incidence* and the *angle of reflection.*

4. Apply the same method that you used in steps 1–3 to construct angles of incidence and reflection from the toes of Polygon's own shoes to the toes of her shoes in the mirror image. How do these angles compare?

5. Use congruent triangles to show the relation between the height of the original character and the height of the mirror image.

Discussion and Extension

1. With geometry utility software, use the same method to construct the figure shown at the right. Follow steps *a–f* to guarantee that the mirror shows the entire image of the original figure.

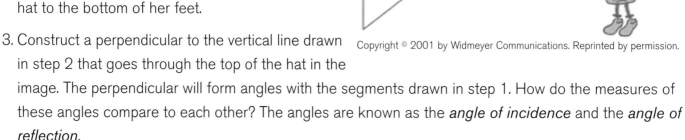

   a. Construct the mirror and label its top *A* and its bottom *B,* as shown.

   b. Choose a point to represent the eye and label it *C*.

   c. Construct the line through the eye (point *C*) that is parallel to the mirror. This line contains the original, which will be specified below in step *f.*

   d. Construct perpendiculars from the top and bottom of the mirror (points *A* and *B*) to the line containing the original figure.

Navigating through Geometry in Grades 9–12

Mirror, Mirror, on the Wall (continued)

Name _____

e. Draw segments from the eye (point *C*) to the top (point *A*) and bottom (point *B*) of the mirror.

f. Reflect the segments drawn in step *e* over the perpendiculars created in step *d*. Find the intersection points with the line containing the eye (point *C*) from step *c*. Let the size of the original figure be determined by the intersection points.

2. Use the geometry software to drag the original back and forth toward the mirror. How does the size of the mirror compare to the original as the original moves toward and away from the mirror?

3. Use the knowledge gained through this activity to estimate the longest possible "full-length" mirror that a store should sell. Provide reasons for your estimate.

Slide Me Now

Name _____

1. Using the drawing below, reflect △ABC in line *m* to obtain △A'B'C'.

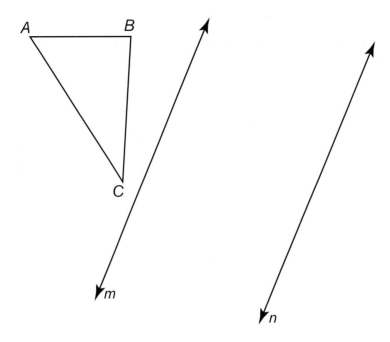

2. Reflect △A'B'C' in line *n* to obtain △A"B"C". (Line *n* is parallel to line *m*.)

3. Measure **AA"**, **BB"**, and **CC"**.

4. Measure the distance between lines *m* and *n* on a perpendicular between the two.

5. Explain why the segments measured in step 3 are parallel and equal in length.

6. Use the information from steps 3 and 5 to argue that △A"B"C" is the translation image of △ABC under the translation that takes A to A".

Discussion and Extension

1. How do your answers in steps 3 and 4 above compare?

2. Given any translation that takes *A* to *B*, find two reflecting lines that could be used to achieve the same result.

Slide Me Now (continued)

Name _____

3. Are the reflecting lines that you found in Discussion and Extension question 2 unique? Why, or why not?

4. Consider the drawing below, and explain how the area of the parallelogram *ABCD* can be found from the area of a rectangle obtained by translating the shaded triangle.

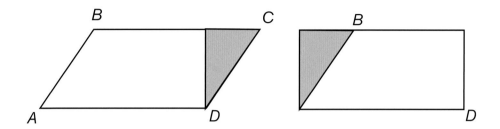

Do you think this idea could be adapted for any parallelogram? Why, or why not?

5. A translation by $\overrightarrow{AB}$ followed by a translation by $\overrightarrow{BC}$ can be accomplished by using four reflecting lines—two parallel lines for each vector. However, the same result can be accomplished using only two parallel reflecting lines. Using the drawing below, determine two parallel reflecting lines that could be used to construct the composite translation.

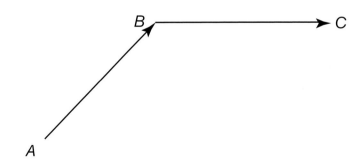

Design This

Name _____

1. Choose and label any point *P* in the top line of the design at the right. For example, let *P* be the upper left tip of the *I* in *Inversions*.

2. The art reads the same right side up or turned around. Find the corresponding image *P'* of point *P* in the inverted image.

3. Repeat steps 1 and 2 for another pair of points, *A* and *A'*.

4. Find the intersection of the segments $\overline{PP'}$ and $\overline{AA'}$, and label this intersection point *O*.

5. Trace the top line of the art on tracing paper. Stick a pushpin at point *O* and turn the tracing paper. Does the traced art match the bottom line of the design? If so, through how many degrees has the tracing been rotated?

6. Use the diagram at the right to reflect △*ABC* in line *m* to obtain △*A'B'C'*.

7. Reflect △*A'B'C'* in line *n* to obtain △*A"B"C"*.

8. Find the perpendicular bisectors of *AA"*, *BB"*, *CC"*.

9. Measure ∠*AOA"*, ∠*BOB"*, ∠*COC"*. Also measure the angle between *m* and *n*.

10. Use the information in steps 8 and 9 to argue that △*A"B"C"* is the image of △*ABC* under the rotation with center *O* through ∠*AOA"*.

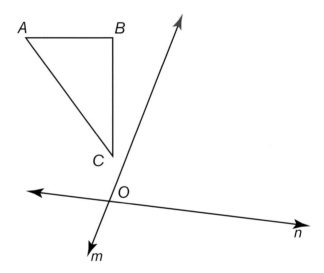

Discussion and Extension

1. Find two reflecting lines that could be used to accomplish the rotation in Scott Kim's logo.

2. Are the lines that you found in Discussion and Extension question 1 unique? Explain your answer.

3. To rotate a figure 270°, what can you say about the angle measure between reflecting lines that could be used?

4. Find reflecting lines that could be used to rotate an equilateral triangle onto itself.

Navigating through Geometry in Grades 9–12

Gliding Along

Name _____

1. In the diagram below, reflect △*ABC* in line *m* to obtain △*A'B'C'*. Next, reflect △*A'B'C'* in line *n* to obtain △*A"B"C"*. Finally, reflect △*A"B"C"* in line *p* to obtain △*A'''B'''C'''*. The resulting transformation is a *glide reflection.*

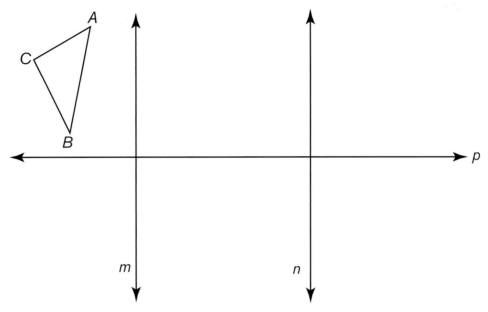

2. Connect points *A, B,* and *C* to their final images, *A'''*, *B'''*, and *C'''*, to form segments. Find the midpoints of these segments. What do you observe? Argue that your observation is true.

3. Given that the two congruent figures shown below are glide reflection images of each other, find three reflecting lines that could be used to make the glide reflection.

Gliding Along (continued)

Name _____

Discussion and Extension

1. Use the reflecting lines given in step 1 of this activity to compose the glide reflection that you obtained there with itself. Describe the resulting transformation.

2. Explain whether the orientation of a figure changes under a glide reflection.

3. In general, would you expect to be able to rewrite the composition of two different glide reflections as a reflection, a rotation, a translation, or another glide reflection? Use orientation in your argument.

4. In the glide reflection, do you think it matters whether the translation is performed first, before the reflection, or whether the reflection is performed first, before the translation? Explain.

5. Explain, giving examples, whether or not the order in which transformations are done matters in general in determining the final image.

Into the Light with Transformations

Name _____

The parabola is a mathematical pattern or shape that can be constructed and represented geometrically, algebraically, graphically, or numerically. To begin, we will construct parabolas geometrically, and then we will extend our exploration to other representational modes.

1. Open a new Cabri Geometry II figure on the TI-92 graphing calculator. Use the **F8/9: Format** option to show the rectangular coordinate axes. Construct the following points, lines, and segments in order and as shown in the calculator screen image below:

   • A segment on the *x*-axis extending from the left edge to the right edge of the screen

   • Point *x* on the segment that lies on the *x*-axis

   • Point *D* on the negative portion of the *y*-axis

   • A line through *D* parallel to the *x*-axis

   • Point *F* in the first quadrant

   • A line through point *x* perpendicular to the *x*-axis

   • Point *B* at the intersection of the lines through *D* and *x*

   • The segment *FB*

   • The perpendicular bisector of *FB*

   • Point *P* at the intersection of the perpendicular bisector of *FB* and the perpendicular line through *x*

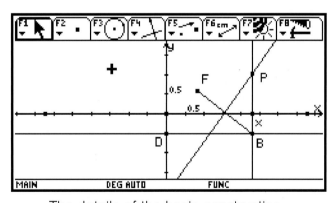

The details of the basic construction

2. Drag point *x* along the *x*-axis and observe how the perpendicular bisector of segment *FB* moves. Make a conjecture about what type of shape this line forms.

3. Select **9:Format** from the **F8** toolbox and set the **Envelope of Lines** option to **OFF** (see screen *a*). Using the **A: Locus** command on **F4** (screen *b*), draw the locus of the perpendicular bisector of *FB* as point *x* moves across its domain (screen *c*). Explain the construction that appears. Drag point *x* and explain what this construction has to do with the perpendicular bisector of segment *FB*.

Into the Light with Transformations (continued)

Name _____

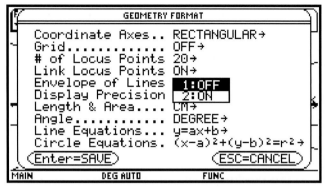

(a) Turn the **Envelope of Lines** off

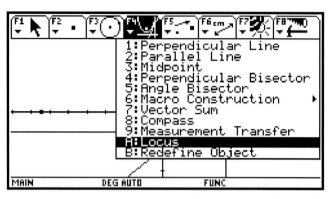

(b) Use the **Locus** command

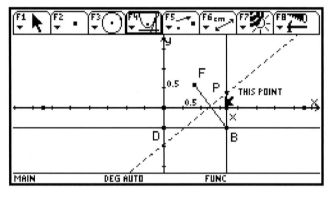

(c) Construct the locus of the perpendicular
bisector of *FB* as *x* moves

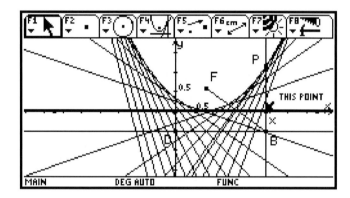

(d) The resulting locus of lines

Drag point *F* around the screen and explain how the figure changes (screen *e*). Explain what happens if point *F* moves below the line through *D* (screen *f*).

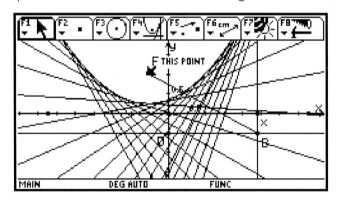

(e) Drag *F* to investigate the shape of the figure

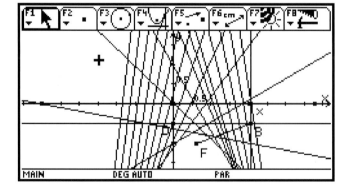

(f) Drag *F* below the line through *D*

4. Using the **Pointer,** select the locus of lines seen in screen *d* and delete it from the screen (with the arrow key ⬅). Use the **A:Locus** tool to construct the locus of point *P* as point *x* moves (screen *g*). Drag point *F* around the screen again, and point *D* along the *y*-axis. Make a conjecture about the shape formed as the locus of point *P* (screen *h*).

Navigating through Geometry in Grades 9–12

Into the Light with Transformations (continued)

Name _____

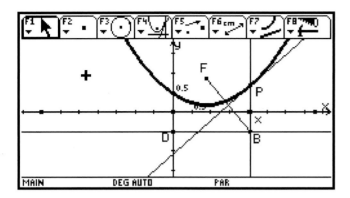

(g) The locus of *P* as *x* moves

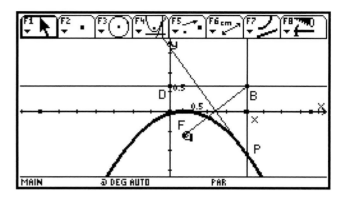

(h) What happens when *F* is below line *DB*?

5. Construct △*FPB* as in screen *i*. Drag point *x* and conjecture about what type of triangle △*FPB* is and how its properties relate point *P* to point *F* and line *DB*. An example is seen in screen *j*. Consult a mathematics text or other source for the geometric definition of a shape that fits this set of specifications. Make a conjecture about the relationship between the perpendicular bisector of segment *FB* and the curve formed as the locus of point *P*. Explain your reasoning geometrically.

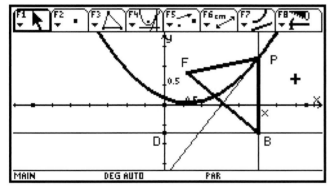

(i) △ *FPB* constructed

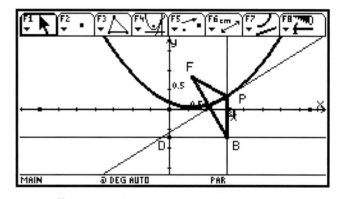

(j) Drag point *x* to see how △ *FPB* changes

6. As point *P* moves along the curve, its coordinates can be transferred to a spreadsheet for analysis. But first, it is necessary to ensure that new data will not be stored with old data in an existing file. The following process will prevent this from occurring. From the **Geometry** screen, press the **2nd** **–** keys in sequence to access the **VAR-LINK** screen of the TI-92. This screen shows all the stored files saved in the memory of the TI-92. In the **MAIN** folder, search for a file named **sysdata**. If you find this file, place the cursor on the file name and press the **··** key (backspace/delete) to erase it from the memory. Press the **ESC** key to return to the **Geometry** screen. (This process must be repeated each time a new set of data is collected.)

Into the Light with Transformations (continued)

Name _____

Use the **5:Equations** and **Coordinates** tool on **F6** to display the coordinates of point *P* (screen *k*). Select the **2:Define Entry** command from the **7:Collect Data** option on **F6** (screen *l*).

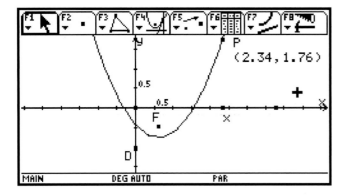

(k) Coordinates of *P* displayed

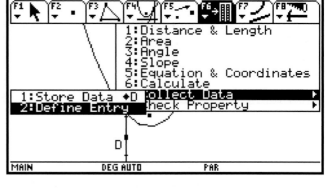

(l) Define the data to be collected

Select the *x*- and then the *y*-coordinate of point *P*, in that order (screen *m*). Move point *x* so that point *P* is near the edge of the screen (screen *n*).

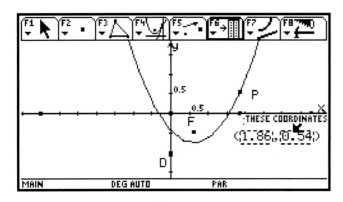

(m) Select the *x*- and *y*-coordinates, in that order

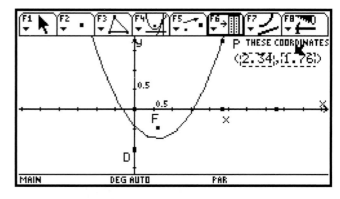

(n) Move *x* so that *P* is near the edge

Use the **1:Store Data** command from the **7:Collect Data** menu to store the current coordinates of point *P* in the data file called **sysdata** (screen *o*). Next, select the **3:Animation** tool from **F7** (screen *p*), and stretch the **Animation** "spring" from point *x* as if you were dragging *x* (screen *q*). As point *x* moves across its domain, the coordinates of point *P* are repeatedly stored in the **sysdata** file. Press the ESC key to stop the data collection after point *x* has traveled across the domain once. To repeat the data collection, delete the **sysdata** file and choose the **1:Store Data** and **3:Animation** steps again (screen *r*). There is no need to **Define Entry** again once this is done.

Into the Light with Transformations (continued)

Name _____

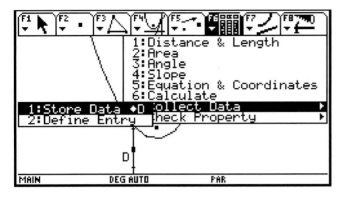

(o) Store the first position of *P* in **sysdata**

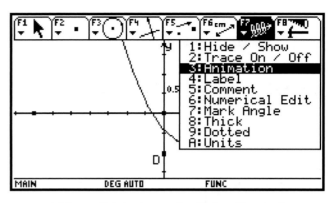

(p) Immediately choose the **Animation** tool

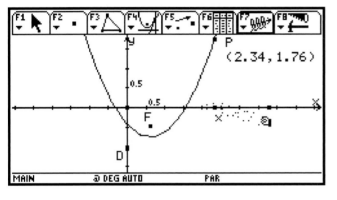

(q) Stretch the **Animation** "spring"

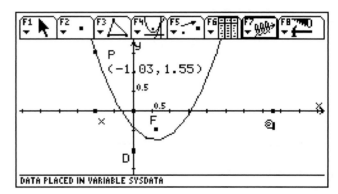

(r) As *x* moves, the coordinates of *P* are stored

7. The collected data can be graphed and analyzed by fitting a curve. Use the APPS key to select the **6:Data/Matrix Editor** (screen *s*). Open the file **sysdata** stored in the main folder (screen *t*).

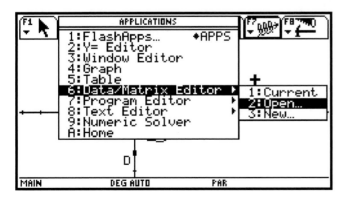

(s) Select the **Data/Matrix Editor**

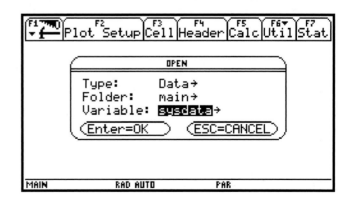

(t) Open the file **sysdata** in the **main** folder

Into the Light with Transformations (continued)

Name _____

Screen *u* shows the file **sysdata,** containing the coordinates of point *P* as it moved along the curve, open in the **Data/Matrix Editor.** (Note that cell width in the **Data/Matrix Editor** can be changed by selecting **9:Format** from the **F1** toolbox.) Choose **F5 Calc** to analyze the data and fit a model. Open the **Calculation Type** menu, and select **9:QuadReg** (quadratic regression) (screen *v*).

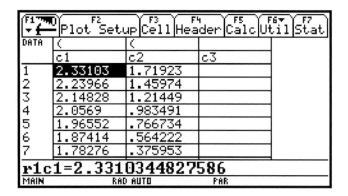

(u) **Sysdata** open in the **Data/Matrix Editor**

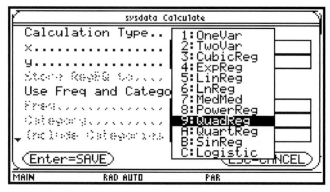

(v) Choose **QuadReg** for the analysis

Enter the names of the columns containing the *x* and *y* data, **c1** and **c2,** and store the regression equation in function **y1(x)** (screen *w*). Screen *x* displays the results of the quadratic regression procedure, showing the three coefficients of the quadratic model and R^2, the coefficient of multiple determination. The value R^2 represents the percent of variability in the data that is explained by this quadratic model. In this case, 100% of the variability is explained. In other words, the model fits the data exactly.

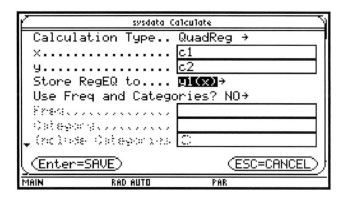

(w) Set up the quadratic regression

(x) The results of the least-squares regression

8. Now that a model has been calculated, the data and the model can be graphed to see if there is a good correlation between the two representations. Screen *y* shows the equations of the model stored on the **Y=** menu and ready to be graphed. Select **Plot 1:** and choose **Scatter** as the **Plot Type**, to make a scatterplot of the data stored in columns **c1** and **c2,** using a **Dot** for each data point (screen *z*).

Navigating through Geometry in Grades 9–12

Into the Light with Transformations (continued)

Name _____

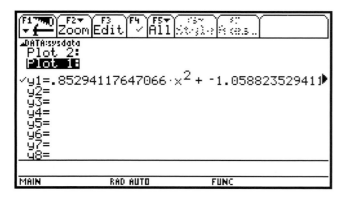

(y) The algebraic model on the $\boxed{\text{Y=}}$ menu

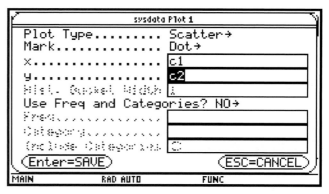

(z) Select **Scatter** to make a scatterplot of the data set

A viewing window appropriate for the graph can be chosen in several ways. Screen *A* shows the selection of the **4:ZoomDec** option from the **F2 Zoom** menu to produce the window seen in screen *B*. The graph in screen *B* has been paused to show the plot of the data set. Screens *C* and *D* show the graph of the algebraic model as it is plotted over the data graph.

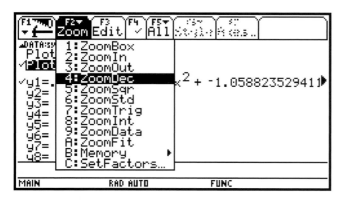

(A) Set a viewing window with **4:ZoomDec**

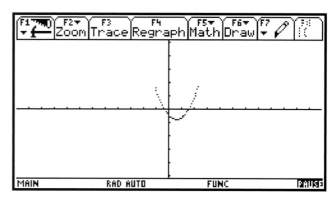

(B) The data set graphed on the screen

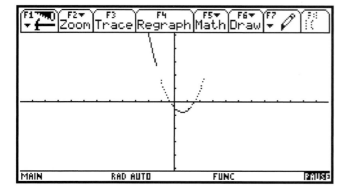

(C) The algebraic model being graphed

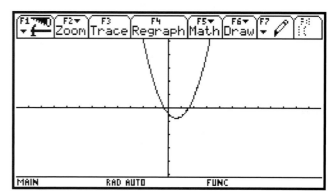

(D) The graph of the data set and the model

Into the Light with Transformations (continued)

Name _____

Discussion and Extension

1. Define a parabola mathematically.

2. Write the general equation of a parabola whose directrix is perpendicular to the *y*-axis.

3. How are the definition and equation of a parabola used throughout this activity?

Transforming with Matrices

Name _____

1. On a sheet of graph paper, plot $\triangle ABC$ with coordinates $A(-5, 1)$, $B(-4, 7)$, and $C(-8, 5)$.

2. On the graph plotted in step 1, draw the line $y = x$.

3. Reflect $\triangle ABC$ in the line drawn in step 2 to find the image $\triangle A'B'C'$. What are the coordinates of A', B', and C'?

4. Translate $\triangle A'B'C'$ by adding 5 to each x-coordinate and leaving the y-coordinates unchanged to find the final image, $\triangle A''B''C''$.

5. Arrange the coordinates of the original triangle as a 3×3 matrix where the last row contains all ones.

6. Use the matrices below for the reflection of step 3 and the translation of step 4:

$$T = \begin{bmatrix} 1 & 0 & 5 \\ 0 & 1 & 0 \\ 0 & 0 & 1 \end{bmatrix}; r_{y=x} = \begin{bmatrix} 0 & 1 & 0 \\ 1 & 0 & 0 \\ 0 & 0 & 1 \end{bmatrix}.$$

Multiply the reflection and translation matrices by the coordinate matrix of step 5 to find the coordinates of the final image triangle for the composition of the reflection and translation, as follows:

$$T \circ r_{y=x}(M) = \begin{bmatrix} 1 & 0 & 5 \\ 0 & 1 & 0 \\ 0 & 0 & 1 \end{bmatrix} \begin{bmatrix} 0 & 1 & 0 \\ 1 & 0 & 0 \\ 0 & 0 & 1 \end{bmatrix} \begin{bmatrix} -5 & -4 & -8 \\ 1 & 7 & 5 \\ 1 & 1 & 1 \end{bmatrix}.$$

7. Do the coordinates obtained in step 6 match the coordinates found in the graph of step 4? Why, or why not?

8. Consider the single matrix obtained by multiplying the three matrices in step 6. What type of transformation does this matrix represent? Explain your answer.

Discussion and Extension

1. Consider the matrix for the reflection given in step 6. Let the matrix

$$\begin{bmatrix} x \\ y \\ 1 \end{bmatrix}$$

represent any point of the plane.

   a. What points are their own images under this reflection?

Transforming with Matrices (continued)

Name _____

b. How do you determine points that are their own images under the reflection in part *a*?

c. Use the matrices and the answer to part *b* to find points that are their own images under the reflection.

d. Are your answers in parts *a* and *c* the same? Why, or why not?

2. Answer Discussion and Extension question 1 using the translation matrix of step 6 above.

3. *Challenge problem:* What type of matrix would you expect to have to write to accomplish a rotation with center (0, 0) and angle of 30°?

Delivering Packages

Name _____

The map on the next page shows the Anoka–Andover–Ramsey–Coon Rapids area of Minnesota. Find the large circled numbers 1–7. The dispatcher for a parcel delivery service must send a driver to these seven locations to deliver packages. Each number identifies one pair of longitude and latitude coordinates from *a–g* below, though not necessarily in order.

a. N 45° 16.146′
      W 93° 24.791′

b. N 45° 14.909′
      W 93° 23.871′

c. N 45° 13.459′
      W 93° 23.047′

d. N 45° 13.239′
      W 93° 21.145′

e. N 45° 14.665′
      W 93°18.938′

f. N 45° 15.732′
      W 93° 18.420′

g. N 45° 14.988′
      W 93° 21.397′

1. Match the correct letter with the corresponding number on the map.

2. Determine the "best" route (that is, the shortest and most direct one) for the driver to follow.

   a. Assume the driver is coming from Highway 10 (at the bottom of the map) and is leaving the highway at one of the exits marked with an *X*.

   b. Keep in mind the driver will have to get back onto Highway 10 in order to return to the warehouse.

   c. Note also that the map shows a river and that there are only three places (marked with an *R*) where the driver can cross it.

   d. Finally, observe that not all straight segments on the map stand for streets; some are grid lines separating the areas of the map. Remember that your driver will not be able to travel on those grid lines.

3. Form small groups to settle on the best route and then trace it on an overhead transparency.

Discussion and Extension:

1. Select a representative who will explain to the class your group's choice of the best route for the driver to follow in delivering the seven packages. Prepare your representative to justify your choice by citing the factors that you took into account and the assumptions that you made.

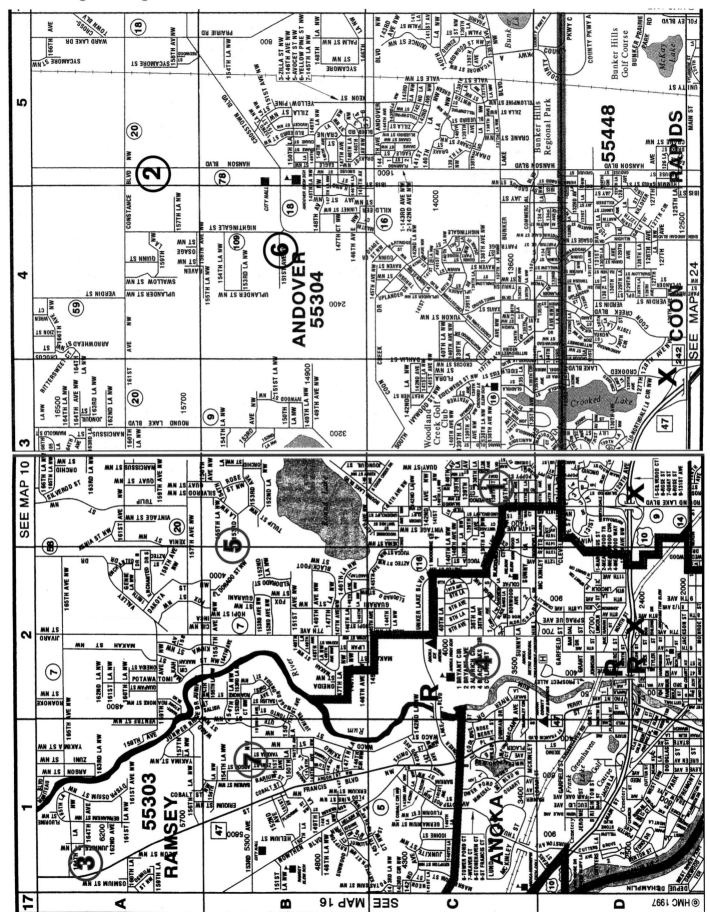

Where Are We Now?

Name _____

Minneapolis–St. Paul International Airport is located at North 44° 50.993′ and West 93°13.030′. San Diego International Airport is located at North 32°43.893′ and West 117°12.131′. During a flight that took off from one of these airports and landed at the other, latitudinal and longitudinal coordinates were recorded at a number of *way points*—places between the major points on a travel route. At the times shown, the following way points were marked:

- At 3:46 Greenwich Mean Time (GMT), N 42°1.675′, W 101°2.590′

- At 4:03 GMT, N 41° 25.599′, W 103° 51.412′

- At 4:20 GMT, N 40°40.125′, W 106°18.641′

- At 5:10 GMT, N 37°45.245′, W 112°12.692′

1. Place the way points as accurately as possible on the map provided on the next page.

2. Examine the way points' coordinates to decide where the plane took off and where it landed. Justify your answer.

Discussion and Extension

1. Would you describe the flight path as linear? Why, or why not?

2. Assuming that the plane's speed stayed approximately constant throughout its flight, calculate approximate times for its takeoff and landing. Explain your reasoning.

3. When the airplane crossed N 43°, what approximately was its western coordinate? At approximately what time did this occur? How do you know?

4. Suppose the clock on the plane's GPS receiver malfunctioned and stayed out of sync with the atomic clock aboard the GPS satellites by .001 second. What effect would this difference have on the accuracy of the receiver's information on the plane's position? Why?

Name _____

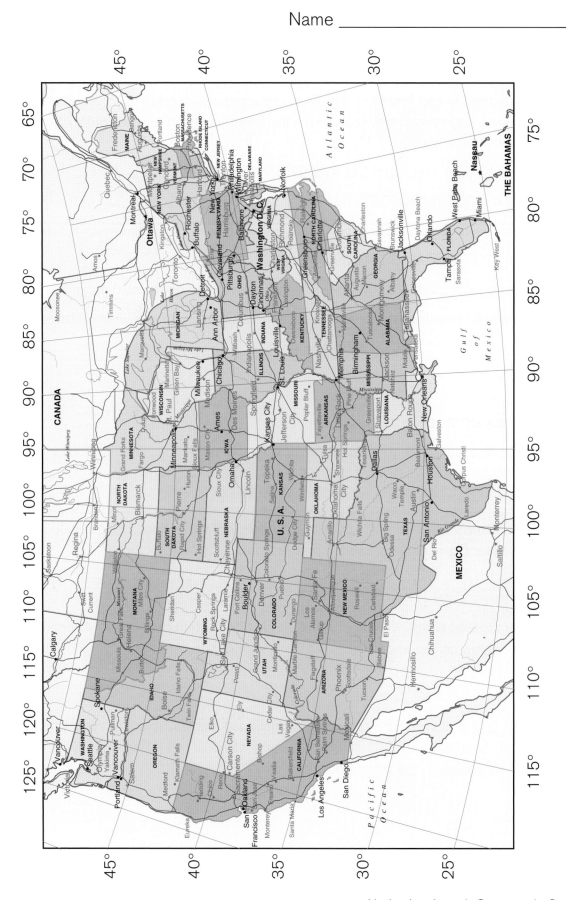

Navigating through Geometry in Grades 9–12

Intuitive Cartography

Name _____

Task *A*

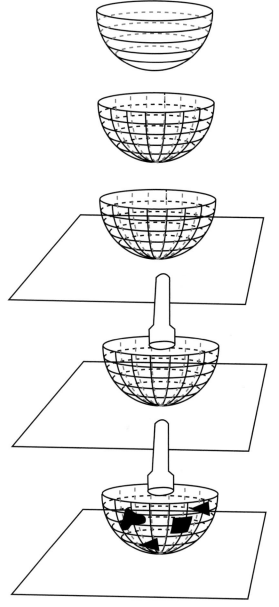

1. On the inside of a plastic hemisphere, draw circles with a colored marker to represent latitudes on Earth. Let the equator be represented by the edge of the hemisphere. Draw three or four latitudes, spaced as evenly as you can make them.

2. On the inside of the hemisphere, draw semicircles, each of which is half of a great circle going through the "pole" of the hemisphere. These semicircles will represent longitudes on the earth. Draw three to five such semicircles, spaced as evenly as you can make them.

3. Place the pole of the hemisphere tangent to a sheet of white unlined paper.

4. With the room darkened, shine a flashlight from the "center of the earth" so that the latitudes and longitudes are projected onto the paper. Trace the projections on the paper. Turn off the flashlight.

5. Using a marker in a different color, draw several shapes on the hemisphere, including at least one that resembles a regular polygon, at least one that looks like an irregular land mass, and at least two that appear to be congruent. Be sure that some of your shapes are near the equator and others are near the pole.

6. Darken the room again, shine the flashlight, and trace the projections of the shapes onto the paper.

Discussion and Extension

1. Describe what happened to the latitudes and longitudes when they were projected onto the paper.

2. Describe what happened to the shapes when they were projected onto the paper.

Intuitive Cartography (continued)

Name _____

3. Roll the hemisphere on the paper to see if the drawings on the hemisphere are the same size and shape as the drawings on the paper. Do they match? Why, or why not?

4. What happened to the shape of the "regular" polygon when it was projected?

5. What shape would you need to draw on the hemisphere near the equator in order to have it projected as a square on the paper? What shape would you need to draw near the pole for a square to appear? Draw examples of those shapes.

6. Discuss which of the following properties you think were preserved in this projection:

   a. Area
   b. "Betweenness" of points
   c. Lines (if you consider a longitude to be a line)
   d. Perpendicularity of lines
   e. Parallel lines
   f. Angle measure
   g. Congruence

Intuitive Cartography (continued)

Name _____

Task *B*

1. Let a spherical light bulb represent the earth. With a colored marker, draw three or four circles on the bulb, spaced as evenly as you can make them, to represent latitudes on Earth. Draw the equator.

2. Draw three to five evenly spaces lines of longitude going through the "poles" of the light bulb. Put the bulb into a lamp base and plug it into an outlet.

3. Place the "North Pole" of the bulb tangent to a whiteboard or chalkboard. Darken the room and turn on the lamp.

4. Trace the projections of the latitudes and longitudes on the board with chalk or a marker. Turn off the lamp.

5. Using a marker in a different color, draw several shapes on the bulb, including at least one that resembles a regular polygon, at least one that looks like an irregular land mass, and at least two that appear to be congruent. Be sure that some shapes are near the equator and others are near the pole.

6. Reposition the lamp against the board as before, turn it on, and trace the projections of the shapes onto it. Turn off the lamp.

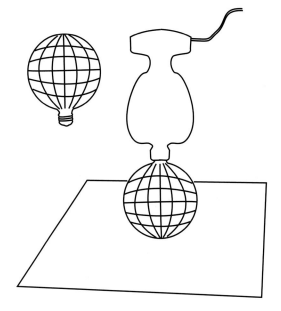

Discussion and Extension

1. Describe what happened to the latitude and longitude lines as they were projected onto the board.

2. Describe what happened to the shapes as they were projected onto the board.

3. Reposition the lamp against the board again, and turn it on. Roll the lamp on the board to see how closely the drawings on the bulb approximate the images on the paper. Do they match? Why, or why not?

Intuitive Cartography (continued)

Name _____

4. What happened to the shape that you intended as a regular polygon when it was projected?

5. What shape would you need to draw on the bulb near its equator in order for a square to be projected on the board? What shape would you need to draw near the pole for a square to appear on the board? Draw examples of those shapes.

6. Discuss which of the following properties you think were preserved in this projection:

   a   Area
   b. "Betweenness" of points
   c. Lines (if you consider a longitude to be a line)
   d. Perpendicularity of lines
   e. Parallel lines
   f. Angle measure
   g. Congruence

Projecting on a Cylinder

Name _____

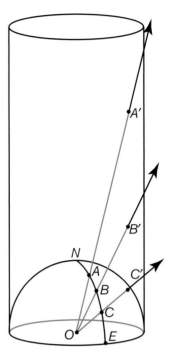

1. Working with a styrofoam hemisphere, find the center of the circle forming its base. With a marker, label the center *O*. Let the outer edge of the hemisphere represent the equator.

2. Label the "North Pole" of the hemisphere as *N*. Label a point on the equator as *E*.

3. Label three equally spaced points *A, B,* and *C* along the longitude containing *N* and *E*.

4. Draw latitudes on the hemisphere.

5. Draw semicircles that go through the hemisphere's pole to represent longitudes.

6. Draw different shapes on the hemisphere, including at least one pair of congruent shapes, one shape near the hemisphere's equator, one near its pole, and one that is intended to be a regular polygon.

7. Wrap a sheet of paper around the hemisphere as shown, making a right circular cylinder that fits over the hemisphere and is tangent to the equator. Place a mark on the bottom edge of the cylinder at point *E* and label it *E'*. Keep *E* and *E'* aligned throughout the activity.

8. Pass skewers through *O* and marked points or vertices on the surface of the hemisphere to find their corresponding points projected on the cylinder. Mark the points on the paper cylinder.

9. Open the cylinder and lay it flat to show the map produced.

Discussion and Extension

1. What happens to points close to *N* on your map?

2. What is the image of point *N* in this projection?

3. Suppose that you had projected the entire sphere onto the cylinder. What shapes on the sphere would be represented by lines on the cylinder?

This activity is based on a module, "What Shape Is Your World?" from *SIMMS Integrated Mathematics: A Modeling Approach Using Technology,* Level 6, Vol.1, developed by the Montana Council of Teachers of Mathematics (1998).

Projecting on a Cylinder (continued)

Name _____

4. Explain which of the following properties are preserved under this projection:

    a. Area

    b. "Betweenness" of points

    c. Lines (if you consider a longitude to be a line)

    d. Perpendicularity of lines

    e. Parallel lines

    f. Angle measure

    g. Congruence

Scale Factors

Name _____

1. On a sheet of paper, use a ruler to help you draw a scalene triangle. (Make it fairly large so that you will still be able to see it well after you reduce it twice on a copy machine!) Label its vertices *A, B,* and *C.*

2. On a copy machine, use a reduction setting of 60% to create a new, smaller triangle from △*ABC.* (Remember, with this setting, you will be reducing your original image by 40% while making a new image that is 60% as big.) Label the new triangle's vertices *A′, B′,* and *C′* to correspond with *A, B,* and *C,* respectively.

3. Repeat step 2 with your new, copied triangle, reducing it also with a setting of 60%. Label this third triangle's vertices *A″, B″,* and *C″* to correspond with *A′, B′,* and *C′,* respectively.

4. Measure the sides of the three triangles. Record their lengths in the table shown here:

Side Measurements

	AB	A′B′	A″B″
Length			
	BC	B′C′	B″C″
Length			
	AC	A′C′	A″C″
Length			

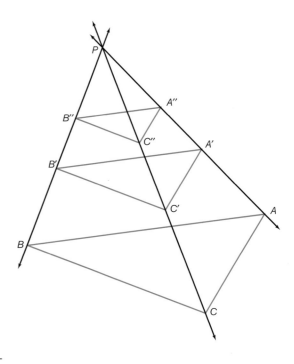

5. Cut out the three triangles. On a new sheet of paper, arrange them from smallest to largest so that corresponding vertices lie along the same line. (That is, vertices *A, A′,* and *A″* should be on a line, *B, B′,* and *B″* should be on a line, and *C, C′,* and *C″* should be on a line.) Use a glue stick to fix the triangles firmly in place, and draw the three lines, extending them across your paper.

6. If you correctly reduced and arranged the triangles, the three lines should meet at a common point. Label this point *P.*

7. Measure and record the segment lengths *AP, A′P,* and *A″P* in the table below:

Segment Length Measurements

Segment	AP	A′P	A″P
Length			

Scale Factors (continued)

Name _____

8. Calculate and compare various ratios of the measurements you have made. Use the following table to record your calculations:

Ratio	Value	Ratio	Value	Ratio	Value	Ratio	Value
A'P/AP		A'B'/AB		B'C'/BC		A'C'/AC	
A''P/A'P		A''B''/A'B'		B''C''/B'C'		A''C''/A'C'	
A''P/AP		A''B''/AB		B''C''/BC		A''C''/AC	

9. Analyze the ratios calculated in the previous step. How do they compare to the copy machine's 60% reduction setting? Summarize your findings in a few sentences.

Discussion and Extension

1. How did the size and shape of triangles *ABC, A'B'C',* and *A''B''C''* change as a result of the copy-machine reductions?

2.   If we repeat this activity using an *enlargement* setting on the copy machine, how will the results change? Present your conjectures in a few sentences. Use the copy machine to check your predictions.

3. In step 5, you drew three lines that were *concurrent.* (Concurrent lines have a point in common.) Do you think you would always find concurrent lines if you reduced and arranged three triangles in this way? Why, or why not?

4. A student made the triangles shown on the next page for an activity like this one. However, the student used a different reduction setting on the copy machine. What setting was it? How did you determine this? Explain your reasoning in a few sentences.

Scale Factors (continued)

Name _____

5. Investigate the final image that results from copying a figure on the copy machine first using a reduction setting of 60% and then using a setting of 75% to reduce the image obtained. When you compare the final image to the original figure, how can you describe the net result of the reduction process?

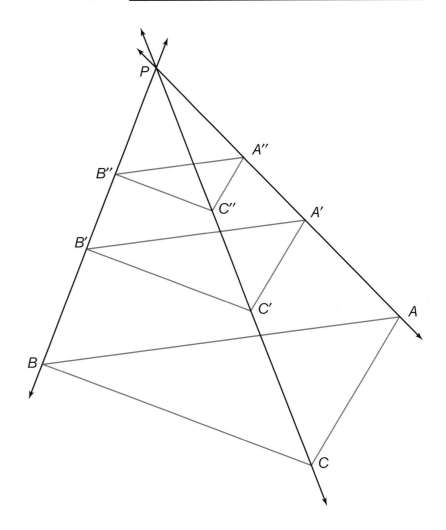

6. Would a reduction made with a setting of 60% followed by a reduction of the image made with a setting of 75% give the same final result as a reduction made with a setting of 75% followed by a reduction of the image made with a setting of 60%?

7. What dilations could you combine to get a final image of each of the following sizes:

   a. 27% of the original
   b. 185% of the original

Basic Dilations

Name _____

1. Quadrilateral *CHOU* has been dilated to generate the similar quadrilateral *KATE.* Determine the scale factor *r* and locate point *P,* the center of the dilation. Write a sentence to describe how you determined *r* and *P.*

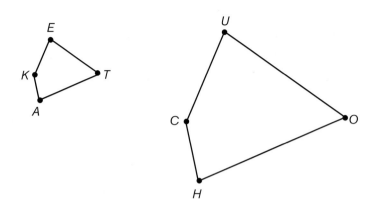

2. Locate an image *BEN* as the dilation of *ART* through point *C* with scale factor 0.50. Now locate an image *LIU* as the dilation of *ART* through point *D* with scale factor 1.25. Both *BEN* and *LIU* are similar to *ART.*

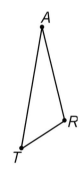

3. Referring to △*ART* in step 2, describe the location of △*SOY* under a dilation of △*ART* through point *D* with a scale factor of 1.

Discussion and Extension

1. Determine a dilation $D_{P,r}$ that maps △*LIU* from step 2 of the activity to △*BEN* (from the same step), identifying *P* and *r* as you progress. If this dilation is not possible to determine, explain why not.

2. Describe the general result of $D_{P,1}$ for any preimage in the plane and center of dilation *P.*

3. In the dilations examined thus far, the center of dilation has always been outside the polygon under transformation. In a triangle, what is the result of a dilation carried out with the center *within* the preimage?

Coordinate Connections

Name _____

1. In the diagram, △*RIM* is a dilation of △*POW* through the origin with a scale factor of 1/3. How do the coordinates of the vertices of the image relate to the coordinates of the vertices of the original triangle?

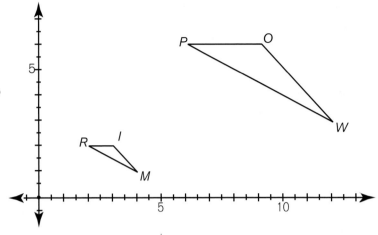

2. Without using a ruler or any other measurement tool, locate a quadrilateral *FARM* in the grid shown so that *FARM* is the image of quadrilateral *LIVE* dilated through the origin with a scale factor of 2.5.

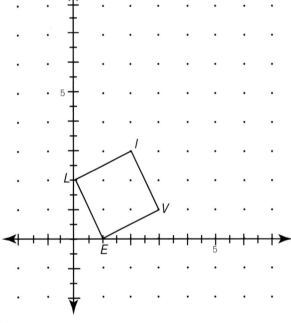

3. The center of dilation *C* of △*WHY* is located at (4, −2). The result of the dilation is the image *NOT*. Determine the scale factor of the dilation.

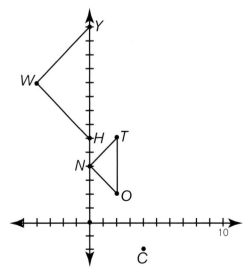

Coordinate Connections (continued)

Name _____

4. The quadrilateral *PLAN* is the image of a dilation of a quadrilateral *TRIM* with center $C =$ (−3, 5) and a scale factor of 0.50. Determine the location of the preimage quadrilateral *TRIM*.

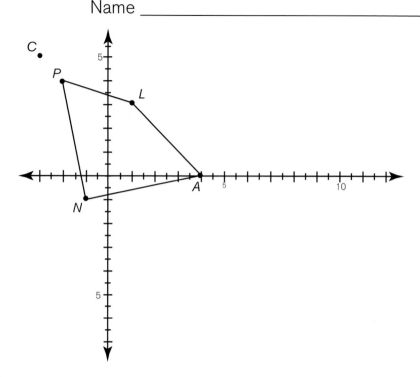

Discussion and Extension

1. Suppose $\triangle ABC$ has coordinates (x_1, y_1), (x_2, y_2), and (x_3, y_3), that the center of a dilation of $\triangle ABC$ is the origin, and that the scale factor is r. Determine the coordinates of $\triangle A'B'C'$, the image of this dilation.

2. Repeat Discussion and Extension question 1 with the center of dilation at the point (c, d).

3. What happens when the scale factor r is negative? Explore the result of a dilation carried out with negative scale-factor values. Describe similarities and differences between dilations with negative scale factors and dilations with positive scale factors.

Multiple Transformations

Name _____

1. Rotate quadrilateral *ABCD* shown at the right 90˚ counterclockwise through the origin to generate the image quadrilateral *EFGH*. Now dilate image *EFGH* through point *P* = (0, 4) using scale factor 3/4 to generate quadrilateral *IJKL*. Locate and identify the final image *IJKL* by the coordinates of its vertices.

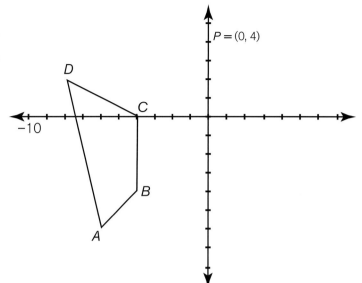

2. Image *XYZ* shown at the right is the result of a dilation and an isometry. Locate the original triangle *RST* that was reflected about the *x*-axis before its reflection image was dilated by a scale factor of 2 through the origin.

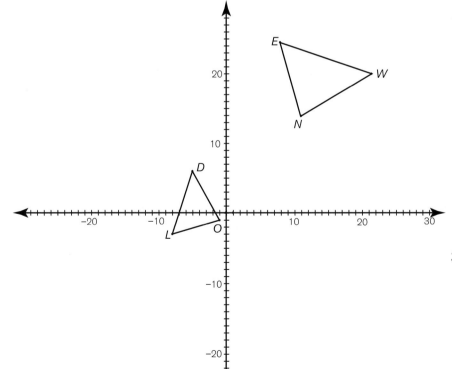

3. Using the two similar triangles *OLD* and *NEW* shown in the coordinate plane at the left, carry out and then describe a composite transformation, including a dilation and an isometry, that will generate the image *NEW* from △*OLD*.

Multiple Transformations (continued)

Name _____

Discussion and Extension

1. How many unique solutions (composite transformations) will produce the image *NEW* in step 3? Explain.

2. Using graph paper, consider a quadrilateral *MATH* that has coordinates (0, 0), (6, 0), (8, 4), and (0, 4). Its image under dilation $D_{P,0.75}$ is the quadrilateral *JOBS*. How does the perimeter of *JOBS* compare with the perimeter of *MATH*?

3. Thinking again about the quadrilaterals *MATH* and *JOBS,* explain how the area of *MATH* compares with the area of *JOBS*?

4. Generalize your results about perimeter and area from the previous two questions to any polygon and its dilation image in the plane.

5. A photocopy machine has 65% and 78% as preset reduction settings and 129% as a preset enlargement setting. Why might these particular scale factors be built into a copy machine?

6. Design a composite transformation that includes at least one dilation. Write instructions for your composition and prepare a model solution for it. Now exchange your instructions with a classmate and carry out each other's directions. Finally, compare the results with the expected solutions and discuss any differences.

Shadowy Measurements

Name _____

1. Go outside on a sunny day. Find a variety of objects whose heights and shadow lengths you can measure easily. Be sure to measure the objects and their shadow lengths at approximately the same time. (Why is that important?) Record the information in the following table or one that uses a similar format.

Object	Length of Shadow	Height of Object

2. Use your data to create a scatterplot, with "shadow length" along the horizontal axis and "object height" along the vertical axis.

3. Study the table and your scatterplot. What relationships do you see? Write an equation that relates shadow length to object height.

4. Now measure the length of the shadow of a nearby building or tree. Use the shadow's measurement and either the scatterplot or the equation to estimate the height of the object.

Discussion and Extension

1. Write a brief report on your work and your findings. Describe the part that similar triangles play in the process of determining an object's height from its shadow's length.

Shadowy Measurements (continued)

Name _____

2. Suppose that on a visit to Washington, D.C., you are standing in view of the Washington Monument. You hold a 6-inch pencil upright 2 feet in front of your eyes, and it exactly covers the monument. If the Washington Monument is 555 feet tall, how far are you standing from its base? Explain the role of similar triangles in your solution.

3. In the third century B.C., the astronomer Eratosthenes used shadows and proportions to estimate the earth's circumference. Study Eratosthenes' work and report to your class how he used shadows and proportions in his calculations. Include in your report a discussion of the accuracy of his estimates.

Field of Vision

Name _____

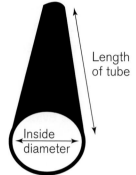

1. Use a viewing tube from your teacher or make a tube from cardboard or construction paper. Measure its length and inside diameter, as shown in the drawing.

2. Look at the table below. You will be working in a small group to gather data to complete it. At the top, record in centimeters the length and inside diameter of your tube.

Tube Length (cm) _____
Tube Diameter (cm) _____

Distance from wall (cm)	Field of vision (cm)
300	
250	
175	
100	
75	
50	
25	
10	
0	

3. Attach a meterstick or measuring tape to the wall at eye level. Select one group member to be the viewer, and position the viewer's eye 3 meters (300 cm) from the wall. Record in centimeters the diameter of the viewer's field of vision through the tube. Continue to measure the viewer's field of vision from each distance indicated in the table from step 2.

4. Use the data collected in the table to create a scatterplot of the ordered pairs (distance from wall, field of vision).

5. Write a paragraph to describe the patterns you detect in your table or on your graph.

6. Repeat steps 1–5 with tubes of different sizes. You may find it useful to vary either the length or the diameter of the tube—not both. Make tables like the one in step 2 to record your data.

Name _____

7. On the basis of your observations, write a conjecture to respond to the following question: For a tube with a given length and diameter, what relationship exists between the viewer's *distance from a wall* and his or her *field of vision?*

8. Provide evidence to justify your conjecture.

Discussion and Extension

Reflect on the solution strategy that you used in step 7 and identify any assumptions or restrictions that you made. How might your solution change if your assumptions changed or your restrictions were relaxed?

What's My Sum?

Name _____

1. Cut a sheet of paper (8 1/2 × 11 in.) into three congruent rectangles, each 8 1/2 inches long.

2. Place two of the rectangles on a table or desk to begin two separate piles, and keep the third rectangle in your hand. Repeat the process with the rectangle in hand: cut it into three congruent rectangles, put two into your piles, and keep one to cut again.

3. Repeat steps 1 and 2 for as long as you can.

Discussion and Extension

1. Imagine repeating the steps above an infinite number of times. How much of the original piece of paper would you have in each pile? In your hand?

2. Assume that the sheet of paper you started with had an area of 1 square unit. Try to write a sum of fractions that represents the part of the paper that you have in either pile.

3. If you were able to continue cutting the paper until you had no paper left in your hand (that is, if you could cut the paper an infinite number of times), would the number of terms in your addition of the fractional parts be finite or infinite? Would the sum itself be finite or infinite?

4. An infinite series is a sum of infinitely many terms. The sum of all the positive integers, $1 + 2 + 3 + 4 + 5 + \ldots$, is an example of an infinite sum. It does not have a "last" number. What do you think this infinite sum should be?

5. A series *diverges* if it does not add up to any finite number. For example, the series that you just considered in Discussion and Extension question 4 diverges. Consider the series $1/2 + (1/2)^2 + (1/2)^3 + (1/2)^4 + \ldots$, formed by adding up all the positive integer powers of one-half. What do you think the answer might be this time? Why? Can you model this with a paper-cutting activity (for example, by cutting a sheet of paper in half, rather than in thirds, and following the process you used before)?

Name _____

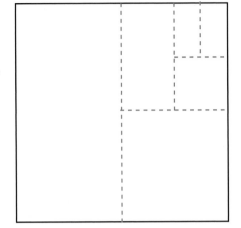

6. Another way to think geometrically about the sum in Discussion and Extension question 5 is to take a square with area 1, as shown, and shade an area that represents the first term (1/2). Now, in the remaining half of that square, shade in an area corresponding to the second term (1/4). Continue shading in this way until you see what the infinite sum might be. What is it?

7. On the basis of this activity, explain whether it is possible for the sum of an infinite series to be a finite number.

Sum Me Up

Name _____

1. Use algebra to divide the polynomial $1 - x^{(n+1)}$ by $1 - x$. Write your answer here:

2. Now suppose that $x = 1/2$. Write the quotient obtained in step 1 as a polynomial in powers of 1/2 by substituting for x.

3. You now have an expression for adding up the first n terms (beginning with the zeroth term) of the series. (The zeroth term, 1, was not included in the infinite series considered in Discussion and Extension questions 5 and 6 in the previous activity, What's My Sum?) What happens to that sum now (with the zeroth term) as n gets infinitely large? Is the sum a finite or an infinite number? What happens to the value of $(1/2)^n$ as n gets infinitely large? Write out some terms (or use a spreadsheet) to answer this question.

4. What is your answer to the original question? That is, what is the sum of the infinite series, $1 + 1/2 + (1/2)^2 + (1/2)^3 + (1/2)^4 + \dots$?

We say that this infinite geometric series *converges* to this sum.

The Koch Snowflake Curve: How Big Am I?

Name _____

1. Set up a spreadsheet with column headings "Number of Iterations," "Number of Sides," "Length of Side," and "Perimeter" (total length), as shown.

	A	B	C	D
1	Number of Iterations	Number of Sides	Length of Side	Perimeter
2				

2. Enter 0 in the A2 cell to represent the initial equilateral triangle with side length 1. Enter 3 in the B2 cell to represent the number of sides of this triangle, and enter a 1 in the C2 cell to represent the length of each side. Finally, enter a 3 in the D2 cell to represent the total length of the sides of the curve.

3. Next, in cell A3, enter the formula "= A2 + 1" to increase the number of iterations by one each time. Use the "fill down" command to fill in all the entries in the column. (About thirty entries should be sufficient.)

4. Next, we investigate the number of sides at each iteration. The first stages of the curve's construction are shown.

At stage 1, each side of the triangle has been divided into thirds, and the middle third has been replaced by two segments that form a triangular wedge. How many sides does the curve then have compared with the previous stage? Write a recursive formula for the number of sides in terms of how many there were at the previous stage. Enter the formula in column *B*. Fill down to complete the column.

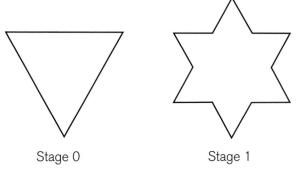

Stage 0   Stage 1

5. Next, consider the length of each side. At stage 0, suppose each of the three sides has length 1 unit. In stage 1, each side is 1/3 as long as at the previous stage. Use the formula for side length to complete column *C*.

6. In column *D*, write a formula that gives the perimeter of the curve by taking the length of each side times the number of sides.

7. Next, use the graphing capabilities of the spreadsheet to produce a graph of the perimeter as a function of the number of iterations.

8. To investigate the area of the snowflake curve, add columns to your spreadsheet with headings "Number of New Triangles" (added in the *n*th iteration), "Area of 1 Triangle," "New Area," and "Total Area," as shown.

This activity includes extensions to three dimensions as described by Collin Joyce when he was a high school student in Freemont, California.

Name _____

	A	B	C	D	E	F	G	H
1	# of Iterations	# of Sides	Length of Side	Perimeter	# of New Triangles Added in nth Iteration	Area of New Triangles	New Area	Total Area
2								

At stage 1, the number of new triangles is 3, each with an area that is 1/9th of that of the original triangle. Determine how the number of new triangles added is derived from the number at the previous iteration. Use this information to help complete column *E*. Next, determine the area of one of the new triangles as a fraction of the area of the original triangle, and use this relationship to fill in column *F*. Multiply the area in column *F* by the number of new triangles in column *E* to fill in column *G*. The total area (column *H)* is the cumulative sum of all the areas.

9. Use the graphing capabilities of the spreadsheet to graph the total area of the curve as a function of the number of iterations.

Discussion and Extension

1. Write a conjecture about what happens to the perimeter of the Koch snowflake curve as the number of iterations increases.

2. Find a pattern in the spreadsheet to help you argue that your conjecture is correct. (*Hint:* Adding a common ratio column may help.)

3. Determine a formula for the area enclosed by the curve at any given iteration *n*. (Use the fact that the sum S_n of the first *n* terms of a geometric series [with first term 1 and ratio *r*] is given by

$$S_n = \frac{1 - r^{(n+1)}}{1 - r}$$

to simplify the formula in row *n* of the last column.)

4. Looking again at the sum in Discussion and Extension question 3, consider what happens as *n* gets very large. What, for example, happens to $r^{(n+1)}$? What happens to the expression for S_n?

5. Imagine a regular tetrahedron with an edge of length *a,* as shown on the next page, as stage 0 of a changing three-dimensional shape exhibiting self-similarity. Suppose that at stage 0, *a* equals 1 unit. Create a spreadsheet with the information indicated. Continue to fill in the spreadsheet as three-dimensional tetrahedra are added to each side in the manner investigated in the snowflake curve activity.

Name _____

The regular tetrahedron shown in (a) has grown a regular tetrahedron (with a side half as long) on each of its faces in (b), and (c) shows this new solid in the process of growing tetrahedra on its faces in the same way.

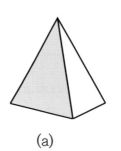
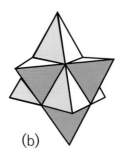
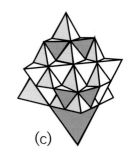

(a)   (b)   (c)

Number of Iterations	Length of a side of a face	Number of faces	Area of each face	Total surface area	# of new tetras	Volume of 1 new tetra	Total new volume	Total volume
0	a	4	$\dfrac{a^2\sqrt{3}}{4}$	$a^2\sqrt{3}$	1	$\dfrac{a^3\sqrt{2}}{12}$	$\dfrac{a^3\sqrt{2}}{12}$	$\dfrac{a^3\sqrt{2}}{12}$
1	$a/2$	$6\times4=24$			4			

6. What is the rate of increase of the total area as more tetrahedra are added?

7. What is the total surface area as more tetrahedra are added? Can you tell?

8. What is the rate of increase of the volume as the number of tetrahedra grows?

9. What formula can you give for the total volume?

10. What would happen if a square were allowed to "grow" small squares on each side?

11. What would be the three-dimensional analog of the figure considered in Discussion and Extension question 10?

12. What would be the analog for any *n*-gon and its three-dimensional counterpart?

Smoke and Mirrors

Name _____

1. Tape two rectangular mirrors together along edges in such a way that the angle between them can be easily changed. You will be making a simple kaleidoscope. Place your joined mirrors on a piece of patterned paper, with the taped edges perpendicular to the paper, as shown.

2. Experiment with different angles between the mirrors. Record how the views change. Write a paragraph explaining what you see.

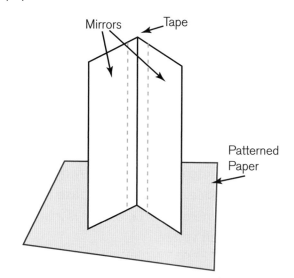

3. Now place your hinged mirrors on a plain sheet of paper. Place a penny or other small object in the space between the two mirrors. Again experiment with different angles between the two mirrors. Write a paragraph about your observations.

4. Experiment with different angles until your mirrors let you see the original object and three images of it in the mirrors. What angle do the mirrors form in this situation?

5. Experiment with mirror angles and make a table to record each angle together with the corresponding number of views of the object (images plus the original) that the angle gives you. How many images plus the original can you possibly see? Do you think that there may be some that you don't see? What angle would you need to make between the mirrors to give yourself eight views of the object (seven images plus the original)?

6. Draw a short line segment with a dot at the end on plain paper. Align the mirrors so that the dot is under the edge of one of them, as shown. Then adjust the mirror angle so that the line segment and its reflections in the mirrors form a regular polygon with 3, 4, 5, 6, …, and up to 12 sides. Record your findings in a table like the one you made in step 5, showing angle measures with corresponding numbers of sides. Can you find two different angles that make the same polygon image? What are those angle measures?

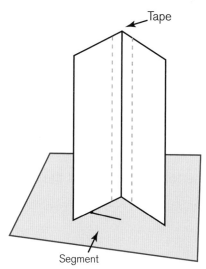

Discussion and Extension

1. Write a paragraph to summarize your findings.

This activity is based on a project found in *Discovering Geometry: An Inductive Approach* by Michael Serra (1997).

*Dis*continuous, That's What You Are!

Name _____

Suppose that a kaleidoscope has been formed from three rectangular mirrors of the same size joined along edges of the same length to form an equilateral triangle at either end. Also suppose that a ray of light enters the kaleidoscope at some point *x* along one mirrored side and travels through the kaleidoscope on a path parallel to one of the other sides, as shown.

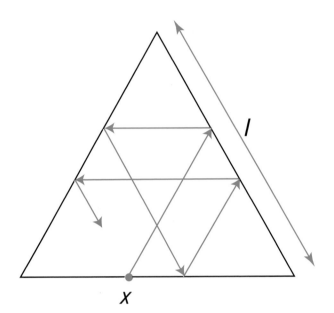

1. With pencil and graph paper or isometric dot paper (or using geometry software), draw an accurate sketch of an equilateral triangle. Position it so that it has a vertex on the lower left that you can label with the coordinates (0, 0).

2. Carefully sketch the path of a ray of light that enters the kaleidoscope at the base of the triangle at some point $P(x, 0)$ that is *x* units from the lower left vertex.

3. Measure the total distance traveled by the light ray from its starting point at $P(x, 0)$ to its arrival back at this point. Be sure to consider a starting point at the midpoint of the triangle.

4. Make a table that shows lengths of paths of light and the corresponding starting values of *x*.

5. Sketch a graph of the total distance traveled by the light (length of the light ray's path) as a function of the distance of its starting point from one of the vertices (you choose it and fix it) on the side where it entered.

Discussion and Extension

1. What is the domain of this function?

2. What is the total distance that the light will travel?

How Small Are the Squares?

Name _____

For this activity, first you need to think of two variables that may have some kind of relationship to each other. For example, you may conjecture that the heights and wrist circumferences of your classmates are related in the sense that as height increases wrist circumference also increases. After you have settled on your variables, collect a small data set.

1. Working with a statistics utility package, make a data table for your variables.

2. Use the table from step 1 to provide the horizontal and vertical components for a graph and make a scatterplot of your data.

3. With the graph displayed, select a movable line and experiment with it until you think it "fits" the data points as closely as possible.

4. Select "show squares" to display the square of the distance from each data point to your movable line. You will notice that the sum of the squares is displayed at the bottom of your screen. Experiment by moving your line, and describe what you see. Is it easier now to guess at where the least squares line of best fit might be? Explain.

5. Set the statistics utility package to display the actual least squares line of best fit. Now you will see your line and the actual "best fit" line simultaneously, as well as the squares of the distances of the data points from these lines. How does the actual line compare to your guess?

Discussion and Extension

Describe the concept of "least squares line of best fit" in a paragraph that would explain it to a friend who was absent from the class when you completed this activity.

Solutions to Blackline Masters

Solutions to Fold Me! Flip Me! (p. 78)

2.

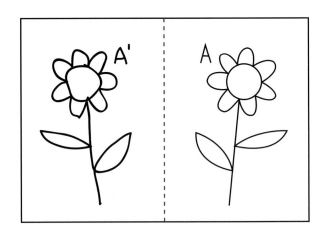

3. The fold line is the perpendicular bisector.

5. *a.* Line *m* is the perpendicular bisector of each segment.

   b. Construct a line through *P* that is perpendicular to the reflecting line. Label the intersection point *X*. Find the point *P'* so that *P'* is on the perpendicular and *PX = XP'*.

   c. Reverse the steps in 5(*b*).

   d. Do the described construction with the compass and straightedge.

6. Draw any two chords and construct the perpendicular bisectors of the chords. The center of the circle is the intersection of the two perpendicular bisectors.

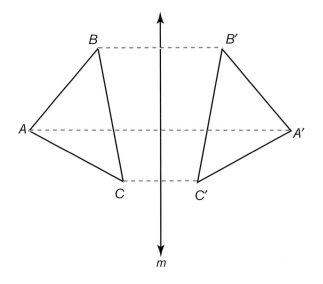

Discussion and Extension

1. The orientation is reversed.

2. The orientation is the same.

3. The orientation changes for an odd number of reflections. It remains the same for an even number.

4. A circle is its own image through a reflection in any line containing the center of the circle.

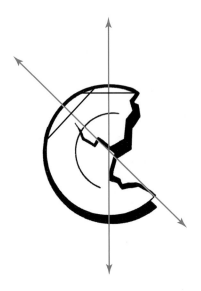

Navigating through Geometry in Grades 9–12

Solutions to Mirror, Mirror, on the Wall (p. 80)

3. The angles have the same measure.

4. The angles have the same measure.

5. The two triangles formed at the top of the drawing are congruent, and the two triangles formed at the bottom of the drawing are also congruent. The distance from head to toe is exactly twice the length of its image in the mirror using the fact that congruent parts of congruent triangles are congruent.

Discussion and Extension

2. The length of the image in the mirror is always half the actual distance from head to foot.

3. Because a full-length mirror image is half the height of an individual, most full-length mirrors are no longer than four feet. Few people are more than eight feet tall.

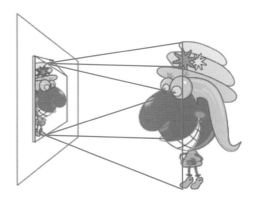

Solutions to Slide Me Now (p. 82)

1–2.   See drawing (a).

5. The segments are perpendicular to the same line and must be parallel. The length of each segment is twice the distance between the parallel reflecting lines; hence they are equal.

6. Each point is "slid" the same distance in the same direction.

Discussion and Extension

1. The distance between the two reflecting lines is half the distance between a point and its image.

2. One reflecting line could be perpendicular to the direction of the translation and through the initial point describing the translation. The other reflecting line could be the perpendicular bisector of the vector marking the translation (see drawing [b]). They are used in the order found here.

3. There are infinitely many other sets of reflecting lines that work, but they are all parallel to the ones identified in Discussion and Extension answer 2, and the distance between any pair used must be half the length of the translation vector.

4. The area of the parallelogram is separated into two parts that can be reassembled to form a rectangle. The area of the parallelogram is the same as the area of the rectangle.

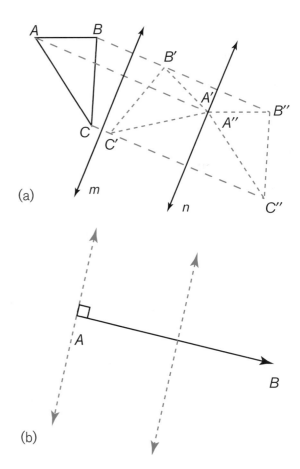

5. The two new lines can be found by applying the ideas in Discussion and Extension answer 2 to the resultant vector of the original two translations (see drawing [c]).

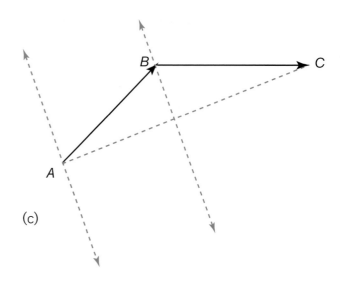

(c)

Solutions to Design This (p. 84)

1–4.

5. Yes; 180°.

7–8. See drawing.

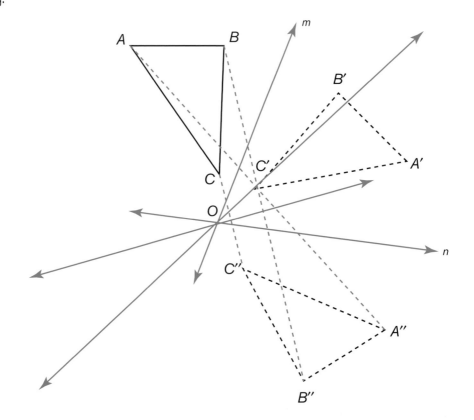

Discussion and Extension

1. Any pair of perpendicular lines through the "center" of the drawing will work.

2. No; see Discussion and Extension answer 1.

3. The measure of the angle should be 135˚.

4. Any reflecting lines that (1) form a 60˚ angle and (2) pass through the "center" of the equilateral triangle will work.

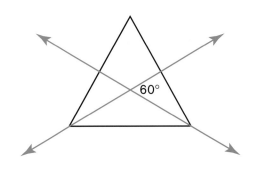

Solutions to Gliding Along (p. 85)

1–2.

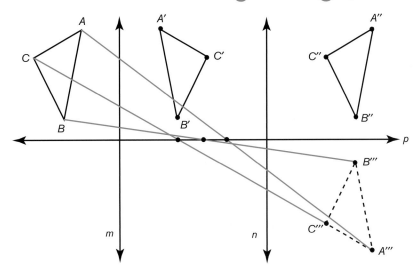

2. The midpoints all lie along the same line. For example, reflecting line *p* is the perpendicular bisector of the segment *A″A‴*. In △*AA″A‴*, *p* is the perpendicular bisector of one side of the triangle and is parallel to side *AA″*. Thus, *p* must pass through the midpoint of side *AA‴*. Similarly, we can show that *p* contains the midpoints of *BB‴* and *CC‴*.

3.

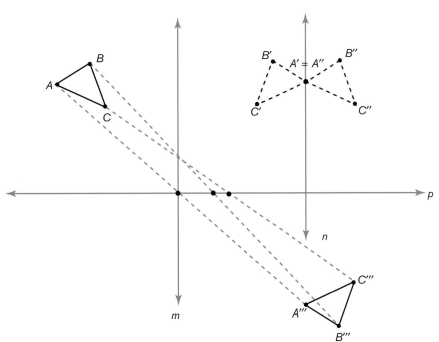

Lines *m, n,* and *p* can be used in that order as reflecting lines.

1. It is a translation.

2. The orientation changes.

3. The composition is either a translation or a rotation depending on whether the reflection lines of the glide reflections are parallel or intersecting. (Use the orientation to show that the composition cannot be a reflection or a glide reflection.)

4. With a glide reflection, it does not matter whether the reflection is performed before the translation or the translation is performed before the reflection. The final image is the same.

5. In general, the order does matter. Consider a translation accomplished by reflecting in two parallel lines m and n. Reflecting in m and then in n results in a very different translation from reflecting in n and then in m. Both result in translations along the same line, but the directions and the lengths of the translations will be different.

Solutions to Into the Light with Transformations (p. 87)

2. A parabola.

3. The line contains a point of the parabola, the point P. The parabola will be "cupped down" when *point F* is located below line *DB*.

4. It is a parabola.

5. Triangle *FPB* is isosceles. Again a parabola is formed, since the distance from the focus F to the point P is the same as the distance from point P to the directrix (line *DB*).

Discussion and Extension

1. A parabola is a set of points in a plane such that all the points are the same distance from a focus and a directrix. (This definition is exemplified in steps 1–5 of the activity.)

2. A parabola has the general equation $y = ax^2 + bx + c$, where a is not equal to 0. (This equation is exemplified in steps 6 and 7 of the activity.)

Solutions to Transforming with Matrices (p. 95)

3. $A'(1, -5)$, $B'(7, -4)$, and $C'(5, -8)$.

4. $A''(6, -5)$, $B''(12, -4)$, and $C''(10, -8)$.

5.
$$\begin{bmatrix} -5 & -4 & -8 \\ 1 & 7 & 5 \\ 1 & 1 & 1 \end{bmatrix}.$$

6.
$$T \circ r \circ M = \begin{bmatrix} 1 & 0 & 5 \\ 0 & 1 & 0 \\ 0 & 0 & 1 \end{bmatrix} \begin{bmatrix} 0 & 1 & 0 \\ 1 & 0 & 0 \\ 0 & 0 & 1 \end{bmatrix} \begin{bmatrix} -5 & -4 & -8 \\ 1 & 7 & 5 \\ 1 & 1 & 1 \end{bmatrix} = \begin{bmatrix} 6 & 12 & 10 \\ -5 & -4 & -8 \\ 1 & 1 & 1 \end{bmatrix}.$$

7. The coordinates match.

8. The composition represents a glide reflection. It is the composition of a reflection and a translation in which the direction of the translation is perpendicular to the reflecting line.

Discussion and Extension

1. *a.* Every point on the reflecting line is its own image.

b. You can solve systems of equations using (x, y) as the coordinates of the points you seek. For points that are their own reflection images,

$$\begin{bmatrix} 0 & 1 & 0 \\ 1 & 0 & 0 \\ 0 & 0 & 1 \end{bmatrix} \begin{bmatrix} x \\ y \\ 1 \end{bmatrix} = \begin{bmatrix} y \\ x \\ 1 \end{bmatrix} = \begin{bmatrix} x \\ y \\ 1 \end{bmatrix}.$$

The multiplication of matrices will find all points (x, y) that are their own reflection image, and they are the points on the reflecting line $y = x$.

c. All points on the line $y = x$.

d. Intuitively, in part 1a, you expect the only points that are their own images under a reflection to be the points on the reflecting line. Now you have discovered that your answers in parts 1a and 1c are the same and consistent with one another.

2. a. There are no points that are their own images.

3. This may be a difficult question for students because trigonometry is required. For advanced students, it is appropriate. The matrix is the following:

$$\begin{bmatrix} \dfrac{\sqrt{3}}{2} & -\dfrac{1}{2} & 0 \\ \dfrac{1}{2} & \dfrac{\sqrt{3}}{2} & 0 \\ 0 & 0 & 1 \end{bmatrix}.$$

Hint: If P is a point with coordinates $(r\cos\theta, r\sin\theta)$, consider the transformation required to locate the image P' with coordinates $(r\cos(\theta + 30°), r\sin(\theta + 30°))$.

Solutions to Delivering Packages (p. 97)

1. Notice the proximity of some of the points, especially those that are similar in their northern coordinates. Look at both the northern and the western coordinates for each point.

   1 (d). This location has approximately the same northern coordinate as (c) but is not as far west.

   2 (f). This location is the farthest east (or the "least west").

   3 (a). This location is the farthest north and the farthest west.

   4 (c). This location has approximately the same northern coordinate as (d) but is farther west.

   5 (g). This location has almost the same northern coordinate as (b) and (e) and lies between them to the east and west.

   6 (e). This location has almost the same northern coordinate as (b) and (g) but is the farthest to the east of the three.

   7 (b). This location has almost the same northern coordinate as (e) and (g) but is the farthest to the west of the three.

Solutions to Where Are We Now? (p. 99)

2. The flight originated at the Minneapolis–St. Paul airport and ended in San Diego. The coordinates decrease in the northerly direction and increase in the westerly direction.

Discussion and Extension

1. The concept of *linear* on a flat map is different from the same concept in spherical geometry. To show the difference, consider a plane's flight from Boston to Tokyo (discussed briefly on p. 40). The part of the earth covered by a line segment drawn between the two cities on most maps is very different from the part covered by a string stretched between the two cities on a globe. Since planes fly in relation to a globe instead of a flat map, it is not surprising that a flight path from Boston to Tokyo may go over the Arctic Circle. The flight from Minneapolis to San Diego, covering much less distance, does not show this difference as clearly.

3. Although it is inaccurate to assume that the plane maintains a constant speed throughout its flight, this assumption allows students to compute approximate times for the plane's takeoff and landing. Using the longitudinal coordinates of the second and third way points, for example, students can subtract to find that the plane traveled approximately 2.5° west in 17 minutes. Since Minneapolis–St. Paul International Airport is about 7° 49′ (or approximately 7.8 degrees) east of the first way point, students can use proportional reasoning to find approximately how many minutes elapsed from takeoff to that point—that is, 2.5 : 7.8 = 17 : *x*. They can calculate that the plane took off about 53 minutes before the first way point was marked, making its takeoff time approximately 2:53 Greenwich Mean Time (GMT). Using the same reasoning, the plane's landing time should be approximately 5:44 GMT, making its total flying time slightly less than three hours. Again, using the time between the second and third way points as a guide, students can surmise that the plane traveled approximately 0° 45′ (or .75 degrees) south for every 2.5° west.

4. Using proportions, we find that when the plane crossed N 43°, the western coordinate was approximately 97° 40′. This happened at about 2:39 GMT.

5. Since the satellite signals travel at the speed of light (approximately 186,000 miles per second), if the clock on the receiver is out of sync by 0.001 second, it could affect the accuracy by approximately 186 miles, rendering the unit essentially worthless. This shows the importance of accuracy in timing.

Solutions to Intuitive Cartography (p. 101)

Task *A*

Discussion and Extension

1. The latitudes are projected as circles. The longitudes become lines.
2. The shapes near the pole are more like their original shapes than the others are.
3. They do not match, although some may be close. Distance and area are not preserved.
4. They no longer appear to be regular.
5. The shape near the equator would need to be more trapezoidal. The sides of the shape near the pole would need to be arcs.
6. *a.* No.
   b. Yes.
   c. Yes.
   d. No.
   e. There are no parallel lines. (Teachers may want to discuss whether or not some arcs on the hemisphere are parallel. What would that mean?)
   f. No.
   g. No.

Task *B*

Discussion and Extension

1. Again, the latitudes become circles, and the longitudes become linear.
2. The shapes near the pole look more like the corresponding original shapes than other shapes do that are farther away.

Navigating through Geometry in Grades 9–12

3. No. Distance is not preserved.

4. It no longer appears to be regular.

5. See answer 5 for Task *A*.

6. See answer 6 for Task *A*.

Solutions to Projecting on a Cylinder (p.105)

Discussion and Extension

1. Points close to *N* go to "infinity."

2. *N* has no image.

3. The lines of longitude.

4. *a.* No.

   b. Yes.

   c. Yes.

   d. Yes.

   e. Lines of longitude become parallel lines.

   f. No.

   g. No.

Solutions to Scale Factors (p.107)

8. Each ratio entered into the first and second rows should be close to 0.60. Each ratio entered into the third row should be close to 0.36.

9. The ratios correspond to the copy machine's 60% reduction setting. The ratios in the first two rows of the table represent one reduction made at a 60% setting. The ratios in the third rows represent a second reduction (a reduction of a reduction) made at a 60% setting.

Discussion and Extension

1. The shape of the triangles does not change. The size changes according to the 60% scale factor. The sides of $\triangle A'B'C'$ are 60% as long as corresponding sides of $\triangle ABC$, the sides of $\triangle A''B''C''$ are 60% as long as corresponding sides of $\triangle A'B'C'$, and the sides of $\triangle A''B''C''$ are 36% (0.6 • 0.6) as long as corresponding side lengths in $\triangle ABC$.

2. The original triangle and the image triangle will have the same shape, as in a reduction, but this image triangle will be larger than the original triangle.

3. Yes, this will always be true, since the reduction or enlargement—any dilation, in fact—emanates from a center point, which is the point of concurrency of the lines.

4. A reduction setting of 75%. This can be determined by examining ratios of sides or distances such as those in the table for activity step 8.

5. The final image will have lengths reduced to 45% of the corresponding lengths in the original figure.

6. Yes, because 0.60 • 0.75 = 0.75 • 0.60.

7. Each combination of dilations must have a product of scale factors (equivalent to reduction settings) equal to the specified final scale factor. For (*a*), two or more dilations whose scale factors have a product of 0.27 will satisfy the requirements. For example, an enlargement to 270% followed by a reduction to 10%, or a reduction to 30% followed by a reduction to 90%. Likewise for (*b*), two or more dilations whose scale factors have a product of 1.85, such as a reduction to 50% (0.5) followed by an enlargement to 370% (3.7).

Solutions to Basic Dilations (p.110)

1. The scale factor $r = 0.40$. It can be determined by calculating one or more ratios of sides of *KATE* to corresponding sides of *CHOU*. Point *P* can be determined by sketching segments that contain corresponding vertices (*EU*, *TO*, *KC*, *AH*). These segments are concurrent at point *P*, the center of dilation.

2. See solution below for Discussion and Extension question 1.

3. *SOY* will be the image of *ART* mapped onto itself because the scale factor is 1. Thus, *SOY* and *ART* will be the same triangle.

Discussion and Extension

1. The scale factor $r = 0.4$ (0.50/1.25). *P* is collinear with *D* and *C*, and $PC/PD = 0.20$.

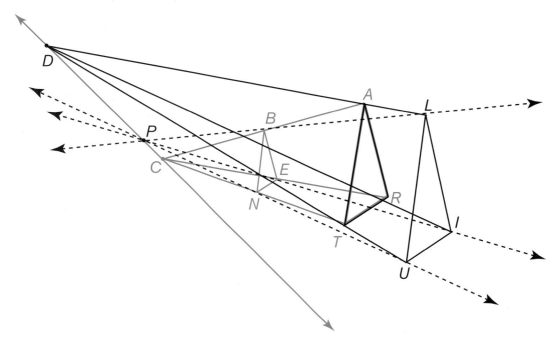

2. A dilation with a scale factor of 1 will always result in an image mapped directly and completely onto the preimage.

3. In a triangle, a dilation carried out with the center within the preimage and the scale factor $0 < r < 1$ results in an image that is entirely contained within the preimage. In a dilation carried out with the center within the preimage and the scale factor $r > 1$, the preimage will be entirely contained within the image. Two examples are shown.

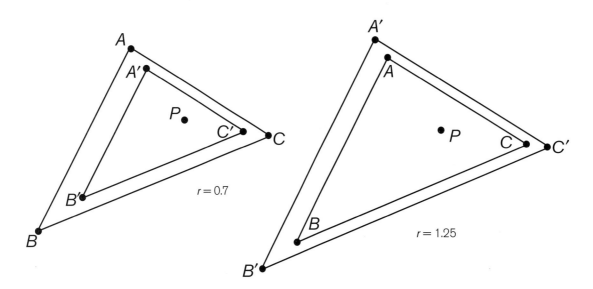

Navigating through Geometry in Grades 9–12

Solutions to Coordinate Connections (p.111)

1. Each of the coordinates of image *RIM* is one-third of the value of the corresponding coordinate of *POW*. For example, point *W* is at (12, 3). The image of *W* under the dilation through the origin is point *M* at (4, 1) since $4 = 1/3 \times 12$ and $1 = 1/3 \times 3$.

2. The coordinates of the vertices of *LIVE* are $L = (0, 2)$, $I = (2, 3)$, $V = (3, 1)$, and $E = (1, 0)$. The corresponding coordinates in the image are $F = (0, 5)$, $A = (5, 7.5)$, $R = (7.5, 2.5)$, and $M = (2.5, 0)$.

3. The scale factor is 0.5. Three ways in which this can be found are by determining (*a*) a ratio of corresponding side lengths, such as *NT* to *WY*, (*b*) a ratio of distances from the dilation center to corresponding vertices, such as *CN* to *CW*, and (*c*) a ratio of change in coordinates from corresponding vertices to the center of the dilation, such as the change in *x* from *N* to *C* compared to the change in *x* from *W* to *C*.

4. *TRIM* has coordinates $T = (-1, 3)$, $R = (5, 1)$, $I = (11, -5)$, and $M = (1, -7)$. The ratio of change in coordinates from corresponding vertices to the center of dilation must be 0.5, the scale factor. We use this relationship with the coordinates of the center and the image points to determine the preimage points. For instance, to determine the coordinates of point *T*, the preimage of *P* under the dilation, we see that the change in the *x*-coordinate from *C* to *P* is 1. The change in the *x*-coordinate from *C* to *T* must be 2, so the *x*-coordinate of *T* must be 2 units to the right of the *x*-coordinate of *C*. The *x*-coordinate of *T* is −1. This same strategy can be employed to determine all desired coordinates.

Discussion and Extension

1. If *r* is the scale factor and $A = (x_1, y_1)$, $B = (x_2, y_2)$, and $C = (x_3, y_3)$, then $A' = (rx_1, ry_1)$, $B' = (rx_2, ry_2)$, and $C' = (rx_3, ry_3)$.

2. If *r* is the scale factor and $A = (x_1, y_1)$, $B = (x_2, y_2)$, and $C = (x_3, y_3)$, and if the center of dilation has coordinates (c, d), then $A' = (r(x_1 - c) + c, r(y_1 - d) + d)$, $B' = (r(x_2 - c) + c, r(y_2 - d) + d)$, and $C' = (r(x_3 - c) + c, r(y_3 - d) + d)$

3. A dilation through center of dilation *P* with scale factor $r < 0$ is equivalent to a 180° rotation through *P* followed by a dilation through *P* with scale factor $|r|$ (i.e., a positive scale factor with the same magnitude as *r*). An example is shown.

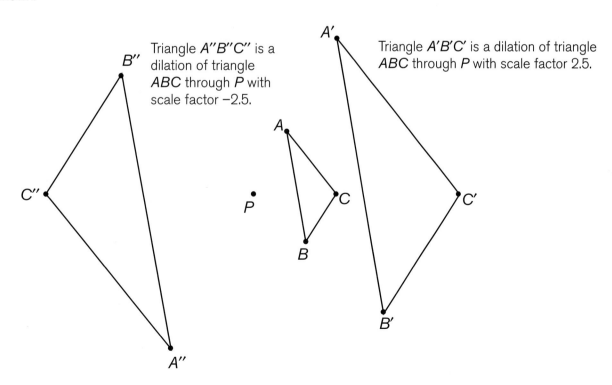

Triangle *A″B″C″* is a dilation of triangle *ABC* through *P* with scale factor −2.5.

Triangle *A′B′C′* is a dilation of triangle *ABC* through *P* with scale factor 2.5.

Solutions to Multiple Transformations (p.113)

1. Quadrilateral *IJKL* has coordinates *I* = (4.5, −3.5), *J* = (3, −2), *K* = (0, −2), and *L* = (−1.5, −5).

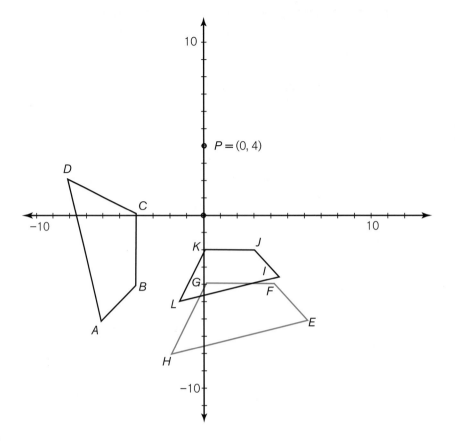

2. Triangle *RST* has coordinates *R* = (0, 4), *S* = (−2, 3), and *T* = (−1, 1). To locate △*RST*, dilate △*XYZ* through the origin with scale factor 0.5 and reflect the image about the *x*-axis.

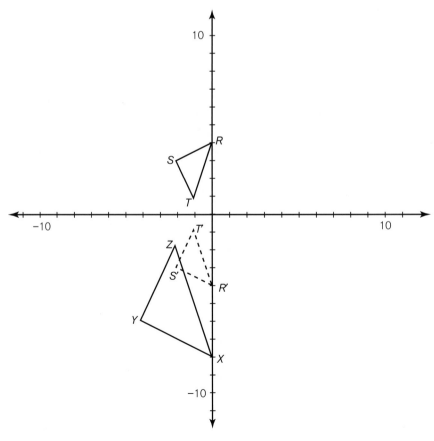

3. A composite transformation that generates
△*NEW* from △*OLD* can be found through the following process. First, we need to determine a rotation angle between a pair of corresponding sides
of the similar triangles. Lines *OQ* and *NQ* can be
used to determine that ∠*OQN* measures 90°. This
angle of rotation can be used in the first transformation. We can rotate △*OLD* 90° clockwise. The
image of this rotation through the origin is
△*O'L'D'*. We now need a dilation that transforms
△*O'L'D'* to △*NEW*. We first determine the scale
factor by comparing lengths of corresponding
sides of the similar figures. This yields a scale factor of 1.5. To determine the center *C* of the dilation of △*O'L'D'* to △*NEW*, we can extend lines
O'N, L'E, and *D'W,* which, we discover, intersect
at the center of dilation *C* at (−25, −25). Refer to
the solution to Discussion and Extension question
1 below for comments on other acceptable transformations.

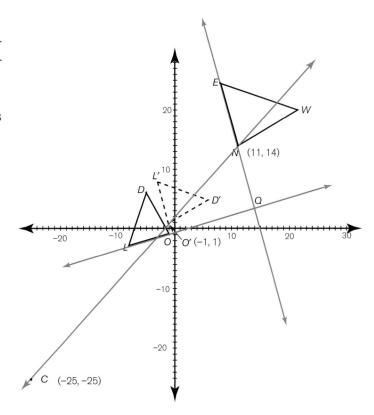

Discussion and Extension

1. There are an infinite number of solutions to problem 3. Two components are required in the composite transformation: (*a*) a rotation of 90° clockwise or 270° counterclockwise (or a multiple of 360° added to either of these)
   and (*b*) a dilation that expands the original image (or some translation or rotation of it) by a scale factor of 1.5.

2. The perimeter of *JOBS* will be 75 percent of the perimeter of *MATH*.

3. The area of *JOBS* will be 0.75 • 0.75 = 0.5625, or 56.25 percent of the area of *MATH*.

4. For dilation scale factor *r* and a polygon with perimeter *p* and area *A,* the perimeter of the image will be *rp,* and
   the area of the image will be r^2A.

5. One possible solution follows: Standard sizes for office paper include 8.5 by 11 inches (letter size), 8.5 by 14
   inches (legal size), and 11 by 17 inches (double letter size). When applied to a standard sheet of paper, each of
   the three preset scale factors results in a new, similar rectangle with a side length that is somehow related to
   other standard paper sizes. For example, a 78% reduction applied to a sheet of legal paper generates a rectangle
   that measures approximately 6.6 by 10.9 inches, thus shrinking the page so that it will fit on letter-size paper. A
   129% enlargement applied to a 78% reduction restores the reduced image to its original size. Other transformations also yield interesting results.

6. Students' responses will vary. See solution to Discussion and Extension question 1 for additional comments.

Solutions to Shadowy Measurements (p.115)

1. See table 3.1 for an example.

2. See figure 3.6 for an example.

3. Equations will vary depending on the time of day, time of year, and place on the earth. All equations should be direct proportions of the form $h = ks$, where s is the object's shadow length, h is the object's height, and k is a constant of proportionality or, equivalently, the slope of the line created in step 2.

4. Answers will vary.

Discussion and Extension

1. The drawings at the right represent the data from table 3.1. The right triangles created by objects and their shadows with the sun's rays are all similar to one another because they share a common angle and all have right angles. Because the triangles are similar, the ratios h_1/s_1, h_2/s_2, h_3/s_3, and h_4/s_4 are equal to one another. This ratio is the slope of the "best fit" line sought in step 3 of the activity.

2. In the drawing below, let E represent the viewer's eye, $\overline{AB}$ be the pencil, and $\overline{MN}$ be the monument. If the viewer moved the pencil out of the way, he or she could look directly ahead at a point P on the monument, as shown. The segment EP, which is equal to the viewer's distance from the monument, is the altitude of $\triangle EMN$ from vertex E to base $\overline{MN}$. Triangle EMN is similar to $\triangle EAB$, since $\overline{AB}$ is parallel to $\overline{MN}$ and $\angle MEN$ is equal to $\angle AEB$. Therefore, the triangles' altitudes from vertex E have the same ratio as their bases, or

$$\frac{2}{.5} = \frac{x}{555} \Rightarrow x = 2220$$

The viewer is 2220 feet from the monument.

$AB = 0.5$ ft.
$ED = 2$ ft.
$MN = 555$ ft.
$EP = ?$

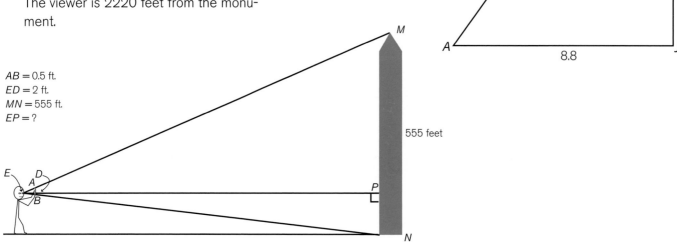

Solution to Field of Vision (p. 117)

The discussion in the section "Peering through a Tube" in chapter 3 provides background information and justification for the results that will emerge from this activity.

Solutions to What's My Sum? (p. 119)

Discussion and Extension

1. You will have one-half of the original piece in each pile and none of it in your hand.

2. $\dfrac{1}{3}+\dfrac{1}{9}+\dfrac{1}{27}+\dfrac{1}{81}+\cdots+\dfrac{1}{3^k}+\cdots$

3. The number of terms is infinite. The sum is finite.

4. Infinite or infinity.

5. $\dfrac{1}{2}+\dfrac{1}{4}+\dfrac{1}{8}+\dfrac{1}{16}+\cdots+\dfrac{1}{2^k}+\cdots=1.$

6.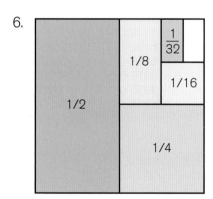

7. Yes, we have seen two examples of infinite series whose sums are finite.

Solutions to Sum Me Up?

1. $$\frac{1-x^{(n+1)}}{1-x}=1+x+x^2+\cdots+x^{(n-2)}+x^{(n-1)}+x^n$$

2. $$1+\frac{1}{2}+\left(\frac{1}{2}\right)^2+\cdots+\left(\frac{1}{2}\right)^{n-2}+\left(\frac{1}{2}\right)^{n-1}+\left(\frac{1}{2}\right)^n, \text{ or}$$

$$1+\frac{1}{2}+\frac{1}{4}+\frac{1}{8}+\frac{1}{16}+\cdots+\left(\frac{1}{2}\right)^n=\frac{1-\left(\frac{1}{2}\right)^{n+1}}{1-\frac{1}{2}}=2\left(1-\left(\frac{1}{2}\right)^{n+1}\right)=2-\left(\frac{1}{2}\right)^n$$

3. The zeroth term adds 1 to the series considered in the previous activity. The sum is a finite number.

   $$\left(\frac{1}{2}\right)^n \to 0 \text{ as } n \to \infty$$

4. 2.

Solutions to The Koch Snowflake Curve:How Big Am I?
(p. 122)

The spreadsheet created by students in steps 1–8 will resemble the following abbreviated sample:

# of Iterations	# of Sides	Length of side	Perimeter	# of New Triangles	Area of 1 Triangle	New Area	Total Area
0	3	1	3	1	0.433012702	0.433012702	0.433012702
1	12	0.333333333	4	3	0.048112522	0.144337567	0.577350269
2	48	0.111111111	5.333333333	12	0.005345836	0.06415003	0.641500299
3	192	0.037037037	7.111111111	48	0.000593982	0.028511124	0.670011424
4	768	0.012345679	9.481481481	192	6.5998E-05	0.012671611	0.682683034
5	3072	0.004115226	12.64197531	768	7.33311E-06	0.005631827	0.688314861
6	12288	0.001371742	16.85596708	3072	8.1479E-07	0.002503034	0.690817896
7	49152	0.000457247	22.47462277	12288	9.05322E-08	0.00111246	0.691930355
8	196608	0.000152416	29.96616369	49152	1.00591E-08	0.000494427	0.692424782
9	786432	5.08053E-05	39.95488493	196608	1.11768E-09	0.000219745	0.692644527
10	3145728	1.69351E-05	53.2731799	786432	1.24187E-10	9.76645E-05	0.692742191
11	12582912	5.64503E-06	71.03090654	3145728	1.37985E-11	4.34064E-05	0.692785598
12	50331648	1.88168E-06	94.70787538	12582912	1.53317E-12	1.92918E-05	0.69280489
13	201326592	6.27225E-07	126.2771672	50331648	1.70352E-13	8.57411E-06	0.692813464
14	805306368	2.09075E-07	168.3695562	201326592	1.8928E-14	3.81072E-06	0.692817274
15	3221225472	6.96917E-08	224.4927416	805306368	2.10311E-15	1.69365E-06	0.692818968
16	12884901888	2.32306E-08	299.3236555	3221225472	2.33679E-16	7.52734E-07	0.692819721
17	51539607552	7.74352E-09	399.0982074	12884901888	2.59644E-17	3.34549E-07	0.692820055
18	2.06158E+11	2.58117E-09	532.1309432	51539607552	2.88493E-18	1.48688E-07	0.692820204
19	8.24634E+11	8.60392E-10	709.5079242	2.06158E+11	3.20548E-19	6.60837E-08	0.69282027
20	3.29853E+12	2.86797E-10	946.0105656	8.24634E+11	3.56164E-20	2.93705E-08	0.6928203
21	1.31941E+13	9.55991E-11	1261.347421	3.29853E+12	3.95738E-21	1.30536E-08	0.692820313

and so on …

Note: The spreadsheet gives approximations; for example, the area in iteration 0 is actually $\sqrt{3}$ divided by 4.

4. The recursive formula for the number of sides of the curve in terms of the number of sides at the previous stage is number of sides = (previous number of sides) • 4.

5–6, 8. Solutions are given in the following equations:

For iteration 0 (original triangle)–
Number of sides = 3
Length of 1 side = 1
Perimeter = (number of sides) • (length of 1 side)
Number of new triangles = 1
Area of 1 triangle = (length of 1 side)^2 • $\sqrt{3}/4$
New area = (number of new triangles) • (area of 1 triangle)
Total area = new area

For iteration 1 (and each successive iteration)–
Number of sides = (previous number of sides) • 4
Length of 1 side = (previous length of side) ÷ 3
Perimeter = (number of sides) • (length of 1 side)
Number of new triangles = (previous number of sides)
Area of 1 triangle = (length of 1 side)^2 • $\sqrt{3}/4$
New area = (number of new triangles) • (area of 1 triangle)
Total area = (previous total area) + (current new area)

7. The graph will resemble the sample at the left below.

8 See the solution for step 5 on the previous page.

9. The graph will resemble the sample at the right below.

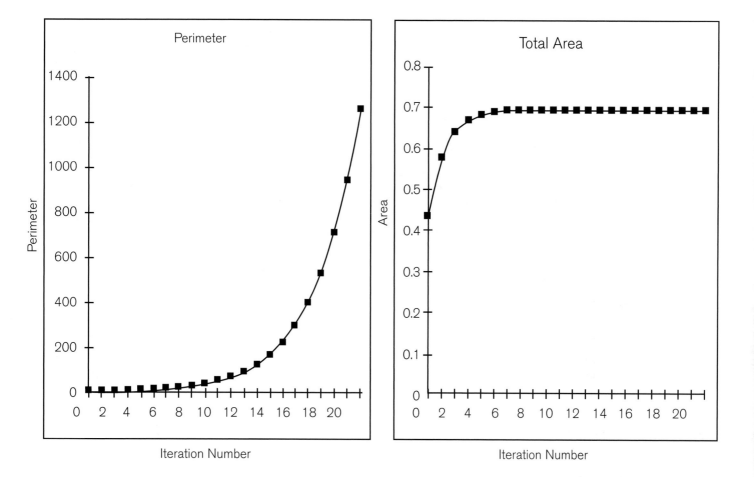

Discussion and Extension

1. Refer to the graph in the solution to step 7 above and the data in the spreadsheet. The perimeter approaches infinity.

2. At every iteration, the perimeter of the Koch curve is equal to the side length times the number of sides. The side length is always 1/3 of the previous length, whereas the number of sides is 4 times its previous number. Thus, the new perimeter at each new iteration is 4 × 1/3, or 4/3, times the previous perimeter. The fact that this number is greater than 1 will support students' conjecture that the perimeter is always increasing and approaches infinity.

3.

$$\frac{\sqrt{3}}{4}\left(1+\frac{3}{4}\left(\frac{4}{9}\right)+\frac{3}{4}\left(\frac{4}{9}\right)^2+\cdots+\frac{3}{4}\left(\frac{4}{9}\right)^n\right)=\frac{\sqrt{3}}{4}\left(1+\frac{3}{4}\left(\frac{1-\left(\frac{4}{9}\right)^{n+1}}{1-\frac{4}{9}}\right)\right)$$

4. As n gets large, $r^{(n+1)}$ approaches 0 if r is less than 1. The expression for S_n becomes

$$\frac{\sqrt{3}}{4}\cdot\left(1+\frac{27}{20}\right).$$

5. The spreadsheet created by the students will resemble the abbreviated sample below:

# of Iteration	Length of Side of Face	# of Faces	Area Each Face	Total Surf. Area	# New Tetras	Vol 1 New Tetra	Total New Vol.	Total Volume
0	1	4	0.433012702	1.732050808	1	0.11785113	0.11785113	0.11785113
1	0.5	24	0.108253175	2.598076211	4	0.014731391	0.058925565	0.176776695
2	0.25	144	0.027063294	3.897114317	24	0.001841424	0.044194174	0.220970869
3	0.125	864	0.006765823	5.845671476	144	2.30178E-04	0.03314563	0.254116499
4	0.0625	5184	0.001691456	8.768507213	864	2.87722E-05	0.024859223	0.278975722
5	0.03125	31104	4.22864E-04	13.15276082	5184	3.59653E-06	0.018644417	0.297620139
6	0.015625	186624	1.05716E-04	19.72914123	31104	4.49566E-07	0.013983313	0.311603452
7	0.0078125	1119744	2.64290E-05	29.59371184	186624	5.61958E-08	0.010487485	0.322090937
8	0.00390625	6718464	6.60725E-06	44.39056777	1119744	7.02447E-09	0.007865613	0.32995655
9	0.001953125	40310784	1.65181E-06	66.58585165	6718464	8.78059E-10	0.00589921	0.33585576
10	0.000976563	241864704	4.12953E-07	99.87877748	40310784	1.09757E-10	0.004424408	0.340280168
11	0.000488281	1451188224	1.03238E-07	149.8181662	241864704	1.37197E-11	0.003318306	0.343598474
12	0.000244141	8707129344	2.58096E-08	224.7272493	1451188224	1.71496E-12	0.002488729	0.346087203
13	0.00012207	52242776064	6.45239E-09	337.090874	8707129344	2.1437E-13	0.001866547	0.34795375
14	6.10352E-05	3.13457E+11	1.6131E-09	505.636311	52242776064	2.67962E-14	0.00139991	0.34935366
15	3.05176E-05	1.88074E+12	4.03275E-10	758.4544665	3.13457E+11	3.34953E-15	0.001049933	0.350403593
16	1.52588E-05	1.12844E+13	1.00819E-10	1137.6817	1.88074E+12	4.18691E-16	0.000787449	0.351191042
17	7.62939E-06	6.77066E+13	2.52047E-11	1706.52255	1.12844E+13	5.23364E-17	0.000590587	0.351781629
18	3.8147E-06	4.0624E+14	6.30116E-12	2559.783824	6.77066E+13	6.54205E-18	0.00044294	0.35222457
19	1.90735E-06	2.43744E+15	1.57529E-12	3839.675736	4.0624E+14	8.17756E-19	0.000332205	0.352556775
20	9.53674E-07	1.46246E+16	3.93823E-13	5759.513605	2.43744E+15	1.0222E-19	0.000249154	0.352805929

and so on …

The formulas for making the spreadsheet are as follows:

For iteration 0 (original tetrahedron):
Length of a side of a face $= 1$
Number of faces $= 4$
Area of each face $=$ (side of face)$^2 \cdot \sqrt{3}/4$
Total surface area $=$ (number of faces) $\cdot$ (area of each face)
Number of new tetrahedra $= 1$
Volume of one new tetrahedron $=$ (side of face)$^3 \cdot \sqrt{2}/12$
Total new volume $=$ (number of new tetrahedra) $\cdot$ (volume of 1 new tetrahedron)
Total volume $=$ total new volume

For iteration 1 (and each successive iteration):
Length of a side of a face $=$ (previous side of a face) $\div 2$
Number of faces $=$ (previous number of faces) $\cdot 6$
Area of each face $=$ (side of face)$^2 \cdot \sqrt{3}/4$
Total surface area $=$ (number of faces) $\cdot$ (area of each face)
Number of new tetrahedra $=$ previous number of faces
Volume of one new tetrahedron $=$ (side of face)$^3 \cdot \sqrt{2}/12$
Total new volume $=$ (number of new tetrahedra) $\cdot$ (volume of one new tetrahedron)
Total volume $=$ (previous total volume) $+$ (current total new volume)

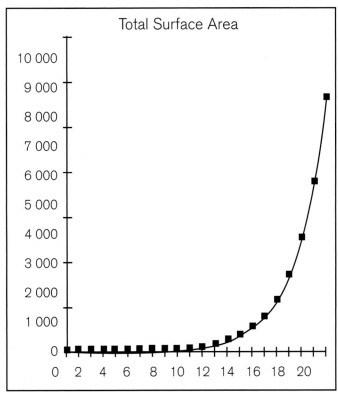

Total Surface Area

Iteration Number

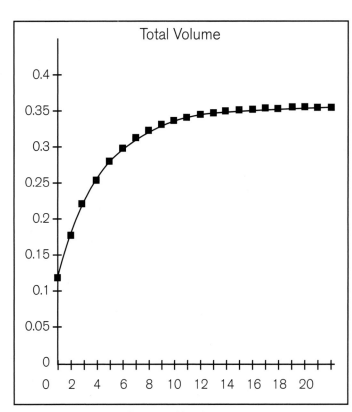

Total Volume

Iteration Number

6. The rate of increase is about 50%, or in other words, the new total surface area is about 150% of the previous surface area.

7. The total surface area continues to grow as new tetrahedra are added. In other words, there is no limit.

8. The rate of increase for the volume goes to zero as the number of tetrahedra grows. That is, there is a limit to the volume ($\sqrt{2}/4$), and the limiting structure of the solid is a cube.

9. The formula for the volume is the following:

$$\frac{\sqrt{2}}{12}+\frac{\sqrt{2}}{8}\left(\frac{3}{4}+\left(\frac{3}{4}\right)^2+\cdots+\left(\frac{3}{4}\right)^n\right).$$

10. If the square grew in a manner comparable to the triangle in Koch's snowflake curve, then the perimeter would grow without bound, and the area of the curve that started with a square of side 1 would approach 2.

11. This would have to be done carefully, but if the cubes of iteration 1 were placed on the sides of the iteration 0 cube so that they occupied the middle part of the original cube, the construction would be possible and would lead to a three-dimensional figure of the Serpinski type.

12. Starting with the pentagon, the added shapes would begin to overlap, as shown. Hence, from this point on, there could be no three-dimensional analogs with the same type of construction.

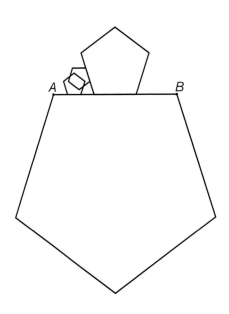

Solutions to Smoke and Mirrors (p. 125)

2. The number of images changes as the angle changes: the smaller the angle, the greater the number of images.

3. Same as for step 2.

4. 90 degrees (or $2\pi/4$).

5. The mirror angle and number of images are related as follows:

Mirror angle	Number of images (not including original)
120 degrees	2
90	3
72	4
60	5
51 3/7	6
45	7
40	8
36	9
$2\pi/n$	$n - 1$

To see seven images, you would have to put the mirrors at an angle of 45 degrees (or $\pi/4$).

6. If the mirrors are placed so that one mirror is perpendicular to the line, then the following results are obtained:

Mirror angle	Number of sides of polygon
60 degrees	3
45	4
36	5
30	6
$180/n$	n

If the mirrors are placed so that they form an isosceles triangle with the line, then the following results are obtained:

Mirror angle	Number of sides of polygon
120 degrees	3
90	4
72	5
60	6
$360/n$	n

Solutions to *Discontinuous, That's What You Are!* (p. 126)

1–5. The length of the path of light from its starting point (on one side of the equilateral triangle representing the kaleidoscope) to its arrival back at that point is $3l$ (where l = the length of a side of the triangle) unless the light starts at the midpoint of a side, in which case the length is $3l/2$. The graph is a discontinuous function with one point of discontinuity.

Discussion and Extension

1. The domain of the function is all numbers x such that $0 \le x \le l$.

2. The distance is either $3l$ or $3l/2$ depending on where the light starts.

Solutions to How Small Are the Squares? (p. 127)

1. The data table should look something like this:

Peopledata

	Name	Height	WristCircumf	\<new\>
1	Bob	75	6.8	
2	Frank	67	6.5	
3	Jane	62	6.1	
4	Sally	64	6.3	
5	Grant	75	6.8	
6	Valerie	61	5.9	
7	Hannah	67	6.3	

2. The scatterplot should look something like this:

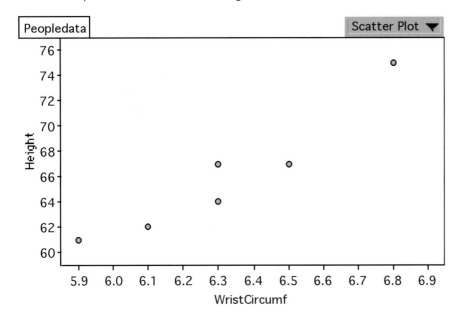

3. The movable line should look something like this:

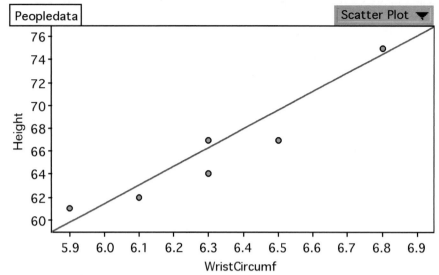

Height = 16.36WristCircumf - 36.7

4. The squares should look something like this:

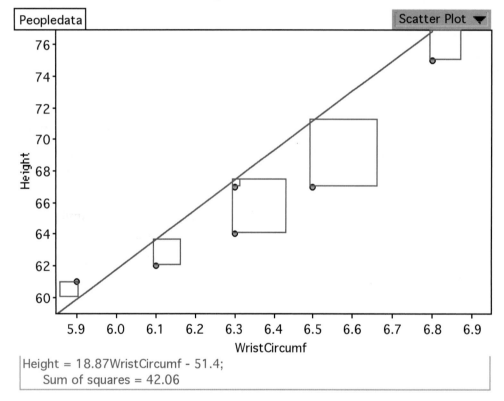

Height = 18.87WristCircumf - 51.4;
Sum of squares = 42.06

The size of the squares changes as you move the line. The sum of the squares also changes. See the bottom of the graph. You can adjust the "movable line" until the sum of squares is as small as possible.

5. The least squares line is shown in purple with the corresponding sum of least squares (for this example) being 14.32. The guess (in this example, the line in black) gave a sum of squares of 42.06.

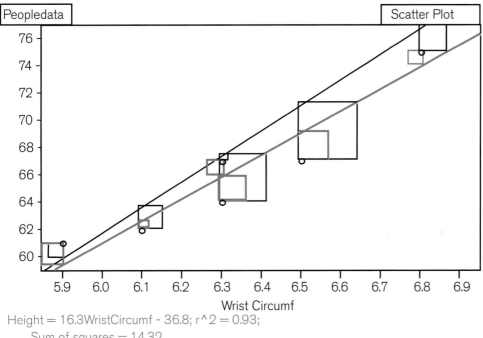

Height = 16.3WristCircumf - 36.8; r^2 = 0.93;
   Sum of squares = 14.32

Height = 18.87WristCircumf − 51.4;
   Sum of Squares = 42.06

6. Answers will vary. A sample paragraph follows:

We measured our wrist circumference and our height. Then we used a dynamic statistics software package to record the data in a table. We then used the "movable line" feature to fit a line (by hand) to our data. I moved it around until I thought it fit the data points quite well. We then used the "show squares" command to show the squares of the distances from the data points to the line we chose. The statistical software package showed the sum of the squares. I then moved the line again until I thought I had it positioned so that the sum of the squares was the least possible. I then used the command "least squares" to show the actual least squares line of best fit for these data. The sum of squares for this line was smaller than my fit. So it really is the best fit!

References

Billstein, Rick, and James Trudnowski. "Godzilla: Fact or Fiction." *NCTM Student Math Notes* (November 1989): 1–4.

Cabri Geometry II. Developed by Yves Baulac, Franck Bellemain, and Jean-Marie Laborde. Austin, Tex.: Texas Instruments, 1998. Software.

Cofman, Judita. *What to Solve? Problems and Suggestions for Young Mathematicians.* Oxford, England: Oxford University Press, 1990.

Coxford, Arthur F., and Zalman Usiskin. *Geometry: A Transformation Approach.* River Forest, Ill.: Laidlaw Brothers, 1971.

Dahl, Bonnie. *The User's Guide to GPS: The Global Positioning System.* Evanston, Ill.: Richardsons' Marine Publishing, 1993.

Dayoub, Iris M., and Johnny W. Lott. *Geometry: Constructions and Transformations.* White Plains, N.Y.: Dale Seymour Publications, 1977.

Eves, Howard. "The History of Geometry." In *Historical Topics for the Mathematics Classroom*, Thirty-First Yearbook of the National Council of Teachers of Mathematics (NCTM), pp. 165–92. Washington, D.C.: NCTM, 1969.

Finzer, William, Tim Erickson, and Lee Binker. "Tour 2: Data, Formulas, and Prediction—Wrist Versus Height." In *Fathom™ Learning Guide*, pp. 23–29. Emeryville, Calif.: Key Curriculum Press, 2000.

Geometer's Sketchpad. Designed by Nicholas Jackiw. Berkeley, Calif.: Key Curriculum Press, 1991. Software.

Haldane, J. B. S. "On Being the Right Size." In *The World of Mathematics*, edited by James R. Newman, vol. 2, pp. 952–57. New York: Simon and Schuster, 1956.

Hoffer, Alan R. *Geometry: A Model of the Universe.* Menlo Park, Calif.: Addison-Wesley, 1979.

House, Peggy. *Mission Mathematics: Linking Aerospace and the NCTM Standards, 9–12.* Reston, Va.: National Council of Teachers of Mathematics, 1997.

Joyce, Collin. "A 3-D Analog to the Snowflake Problem." Freemont, Calif., 2000. Unpublished paper.

Mabry, Rick. "Mathematics without Words." *College Mathematics Journal* 32 (January 2001): 19.

MacPherson, Eric D. "The Themes of Geometry: Design of the Nonformal Geometry Curriculum." In *The Secondary School Mathematics Curriculum*, 1985 Yearbook of the National Council of Teachers of Mathematics (NCTM), edited by Christian R. Hirsch, pp. 65–80. Reston, Va.: NCTM, 1985.

Mandelbrot, Benoit B. *The Fractal Geometry of Nature.* 3rd ed. New York: W. H. Freeman, 1983.

Montana Council of Teachers of Mathematics. "What Shape Is Your World?" *SIMMS Project.* Level 3, vol. 6. Needham Heights, Mass.: Simon & Schuster, 1998.

National Council of Teachers of Mathematics (NCTM). *Principles and Standards for School Mathematics.* Reston, Va.: NCTM, 2000.

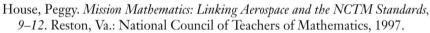

Olson, Alton T. *Mathematics through Paper Folding.* Reston, Va.: National Council of Teachers of Mathematics, 1975.

Papy, Georges. *Modern Mathematics.* Vol. I. London: Collier-Macmillan Limited, 1968.

Pettofrezzo, Anthony J. *Matrices and Transformations.* New York: Dover Publications, 1966.

Phillips, Jo McKeeby, and Russell E. Zwoyer. *Motion Geometry.* Book 1, *Slides, Flips, and Turns.* New York: Harper & Row, 1969.

———. *Motion Geometry.* Book 2, *Congruence.* New York: Harper & Row, 1969.

———. *Motion Geometry.* Book 3, *Symmetry.* New York: Harper & Row, 1969.

———. *Motion Geometry.* Book 4, *Area, Similarity, and Constructions.* New York: Harper & Row, 1969.

Serra, Michael. *Discovering Geometry: An Inductive Approach.* Berkeley, Calif.: Key Curriculum Press, 1997.

Sibley, Thomas Q. *The Geometric Viewpoint: A Survey of Geometries.* Reading, Mass.: Addison-Wesley, 1998.

Snyder, J. P. "Map Projections—A Working Manual." *U.S. Geological Survey Professional Paper 1395.* Washington, D.C.: U.S. Government Printing Office, 1987.

Sobel, Dava. *Longitude: The True Story of a Lone Genius Who Solved the Greatest Scientific Problem of His Time.* New York: Walker, 1995.

Tessellation Exploration. Developed by Kevin Lee. Watertown, Mass.: Tom Snyder Productions, 2001. Software.

Thompson, Richard B. "Global Positioning System: The Mathematics of GPS Receivers." *Mathematics Magazine* 71 (October 1998): 260–69.

Vonder Embse, Charles. "Visualizing Least-Square Lines of Best Fit," *Mathematics Teacher* 90 (May 1997): 404–408.

Wilson, Melvin R., and Barry E. Shealy. "Experiencing Functional Relationships with a Viewing Tube." In *Connecting Mathematics across the Curriculum,* 1995 Yearbook of the National Council of Teachers of Mathematics (NCTM), edited by Peggy A. House, pp. 219–24. Reston, Va.: NCTM, 1995.

Widmeyer-Baker Group. *Figure This! Math Challenges for Families: Take a Challenge, Set 1: Challenges 1–15.* Washington, D.C.: Widmeyer-Baker Group, 1999.

Suggested Reading

Brieske, Tom. "Visual Thinking with Translations, Half-Turns, and Dilations." *Mathematics Teacher* 77 (September 1984): 466–69.

Burke, Maurice. "5-Con Triangles." *NCTM Student Math Notes* (January 1990): 1–4.

Choate, Jonathan, Robert L. Devaney, and Alice Foster. *Fractals: A Tool Kit of Dynamic Activities.* Emeryville, Calif.: Key Curriculum Press, 1999.

Consortium of Mathematics and Its Applications. *Geometry's Future: Conference Proceedings.* Arlington, Mass.: Author, 1990.

Coxeter, H. S. M., and S. L. Greitzer. "The Three Jug Problem." In *Geometry Revisited,* pp. 89–93. Washington, D.C.: Mathematical Association of America, 1967.

Coxford, Arthur F., Jr., Linda Burks, Claudia Giamati, and Joyce Jonik. *Geometry from Multiple Perspectives. Curriculum and Evaluation Standards for School Mathematics* Addenda Series, Grades 9–12, edited by Christian R. Hirsch. Reston, Va.: National Council of Teachers of Mathematics, 1991.

Craine, Timothy V. "Integrating Geometry into the Secondary Mathematics Curriculum." In *The Secondary School Mathematics Curriculum,* 1985 Yearbook of the National Council of Teachers of Mathematics (NCTM), edited by Christian R. Hirsch, pp. 119–33. Reston, Va.: NCTM, 1985.

Devaney, Robert L. *Chaos, Fractals, and Dynamics: Computer Experiments in Mathematics.* Menlo Park, Calif.: Addison-Wesley, 1990.

Egsgard, John. C. "An Interesting Introduction to Sequences and Series." *Mathematics Teacher* 81 (February 1988): 108–11.

Erickson, Tim. "Area and Perimeter: A Study in Limits." In *Data in Depth: Exploring Mathematics with Fathom™*, pp. 111–13. Emeryville, Calif.: Key Curriculum Press 2000.

 Friedlander, Alex, and Glenda Lappan. "Similarity: Investigations at the Middle Grades Level." In *Learning and Teaching Geometry, K–12*, 1987 Yearbook of the National Council of Teachers of Mathematics (NCTM), edited by Mary Montgomery Lindquist, pp. 136–43. Reston, Va.: NCTM, 1987.

King, James R., and Doris Schattschneider, eds. *Geometry Turned On! Dynamic Software in Learning, Teaching, and Research. MAA Notes No. 41.* Washington, D.C.: Mathematical Association of America, 1997.

Larijani, L. Casey. *GPS for Everyone.* New York: American Interface Corporation, 1988.

 Martin, Tami. "Fracturing Our Ideas about Dimension." *NCTM Student Math Notes* (November 1991): 1-4.

Montana Council of Teachers of Mathematics. "One Dish Two Cones." In *SIMMS Project.* Level 3, vol. 3. Needham Heights, Mass.: Simon & Schuster, 1995.

Peitgen, Heinz-Otto F., Hartmut Jurgens, and Dietmar Saupe. *Fractals for the Classroom.* Part 1, *Introduction to Fractals and Chaos.* New York: Springer-Verlag, 1992a.

———. Fractals for the Classroom. Part 2, Complex Systems and Mandelbrot Set. New York: Springer-Verlag, 1992b.

Peitgen, Heinz-Otto F., Hartmut Jurgens, Dietmar Saupe, Evan Maletsky, Terry Perciante, and Lee Yunker. *Fractals for the Classroom: Strategic Activities.* Vol. 1. New York: Springer-Verlag, 1991.

———. *Fractals for the Classroom: Strategic Activities.* Vol. 2. New York: Springer-Verlag, 1992.

———. *Fractals for the Classroom: Strategic Activities.* Vol. 3. New York: Springer-Verlag, 1999.

Wallace, Edward C., and Stephen F. West. *Roads to Geometry.* 2d ed. Upper Saddle River, N.J.: Prentice Hall, 1998.